Creating a Sustainable Healthcare Workplace

Healthcare workers face increasingly dangerous work environments, with patient aggression emerging as a critical threat particularly in emergency departments. This title reveals the alarming frequency of verbal, physical, and mental aggression toward medical staff, often surrounded by a conspiracy of silence despite its significant impact on healthcare delivery. Through research including surveys of 246 medical professionals, 39 interviews, and 13 case studies, the authors expose the prevalence of this phenomenon and its devastating consequences for workers, organizations, and ultimately patient care.

Creating a Sustainable Healthcare Workplace: Management Challenges Related to Aggressive Patient Behaviour identifies, classifies, and addresses workplace threats affecting medical staff safety, with special emphasis on psychosocial hazards related to aggressive patient behaviour. The book presents a three-tiered intervention framework operating at systemic, organizational, and individual levels and offers practical recommendations to prevent incidents, support employees during aggressive encounters, and mitigate aftermath effects. This approach will allow the reader to understand and create a sustainable workplace where healthcare professionals can perform their duties without fear, ultimately improving care quality for all patients.

This essential resource is written for health and safety professionals, healthcare administrators, hospital managers, medical staff supervisors, human resource professionals, and policymakers responsible for healthcare worker safety. Medical professionals working in high-risk environments, particularly emergency departments, will find valuable insights and practical strategies to address patient aggression.

Creating a Sustainable Healthcare Workplace

Management Challenges Related to Aggressive Patient Behaviour

Ewa Beck-Krala, Elżbieta Tarczoń,
Małgorzata Sidor-Rządkowska, and Jerzy Rosiński

CRC Press
Taylor & Francis Group
Boca Raton London New York

CRC Press is an imprint of the
Taylor & Francis Group, an informa business

Designed cover image: Shutterstock

First edition published 2026
by CRC Press
2385 NW Executive Center Drive, Suite 320, Boca Raton FL 33431

and by CRC Press
4 Park Square, Milton Park, Abingdon, Oxon, OX14 4RN

CRC Press is an imprint of Taylor & Francis Group, LLC

ISBN: 978-1-041-19184-1 (hbk)
ISBN: 978-1-041-20466-4 (pbk)
ISBN: 978-1-003-71663-1 (ebk)

DOI: 10.1201/9781003716631

Typeset in Times
by KnowledgeWorks Global Ltd.

Contents

About the authors vii
Foreword ix
Acknowledgment xi

Introduction 1

Chapter 1 Management Challenges in the Healthcare Sector 3

Increasing Demands in Healthcare Work 3
Challenges Facing the Healthcare Sector 4
The Specific Nature of Work in Healthcare 7
Recipients of Medical Services 7
The Intangible Work Environment of Medical Staff 11

Chapter 2 Work Safety as an Element of a Sustainable Healthcare Workplace 13

Sustainable Work Environment 13
The Concept of Occupational Safety 16
Hazards in the Workplace 16
Creating a Culture of Safety in the Workplace 18

Chapter 3 Hazards Occurring in the Healthcare Work Environment 22

Factors Affecting Work Safety 22
Physical Factors in the Work of Medical Personnel 23
Biological Agents in the Work of Medical Personnel 25
Chemical Factors in the Work of Medical Personnel 27
Psychosocial Factors in the Work of Medical Personnel 28

Chapter 4 The Different Faces of Violence and Aggression in Healthcare 34

Aggressive Behavior of Patients 34
Violence and Aggression in the Workplace 37
The Phenomenon of Patient Aggression in the Healthcare Sector 44
The Patient's Perspective and the Phenomenon of Aggression in the Healthcare Sector 46

Chapter 5 Behind the Triage: Research on Patient Aggression in Emergency Departments 49

Survey Among Emergency Department Employees 50
Consequences of Aggressive Behavior by Patients 65

Chapter 6 Multilevel Practical Strategies for Addressing Management Challenges ... 88

Summary of Research Results on Aggressive Behavior of Patients ... 88
Strategies at the Government and Regional Levels ... 94
Strategies at the Organizational Level ... 96
Implementing Change in the Healthcare Sector ... 99
Research Conclusions and Recommendations ... 105

Conclusion ... 119

Appendix ... 123

Bibliography ... 131

Netography ... 141

Glossary ... 145

Index ... 149

About the authors

Ewa Beck-Krala is Professor of Management at AGH University in Krakow, Poland, where she leads the Department of Organization Management and Social Capital. She specializes in Human Resource Management, with research focusing on employee motivation and well-being. She has authored over 100 publications and teaches HR Management, Leadership, and Strategic Management. A Fulbright and Tempus scholar, she has extensive experience in HR Business Partnering and postgraduate education. Her academic and professional expertise spans over 25 years, combining research and teaching with practical insights from business environments.

Elżbieta Tarczoń is affiliated with the 5th Military Hospital in Krakow, Poland, where she specializes in occupational health and safety in healthcare. She holds a PhD in Management and Quality Sciences from AGH University, Poland, and has completed postgraduate studies in occupational health and safety and quality management. Her work focuses on employee safety in healthcare, with published articles on needlestick injuries and preventive measures. She is an active member of the National Association of Occupational Health and Safety Employees and a recipient of the Złote Szelki award for her contributions to workplace safety.

Małgorzata Sidor-Rządkowska is Professor at the Faculty of Management at Warsaw University of Technology, Poland. A graduate of the University of Warsaw, Poland, she transitioned from medical ethics to management sciences to improve healthcare systems. She is an expert for the Polish National Agency for Academic Exchange and a member of the Employers of Poland forum. She has authored 12 books on organizational development and her work focuses on designing healthcare workplaces, informed by her personal connection to the medical profession as the daughter and mother of doctors.

Jerzy Rosiński is Professor of Business Management at Jagiellonian University in Krakow, Poland, and Director of the Institute of Economics, Finance, and Management. His research explores HRM in ICT organizations, digital business models, and teamwork in digital environments. With 25 years of consulting experience, he has led projects such as Talent Bank for the National Bank of Poland and EU Horizon 2020's Una Resin initiative. His work integrates management knowledge with information technology, focusing on competence development and entrepreneurship in the digital age.

About the authors

Foreword

All violence is the result of people deceiving themselves into believing that their pain comes from other people and that, as a consequence, those people deserve to be punished.

Marshall Rosenberg

The growing aggression of patients toward medical staff has become a serious problem in Poland, Western Europe, and the United States. Healthcare workers, including doctors, are increasingly experiencing verbal abuse, threats, and even physical attacks from patients or their families. Such aggression has devastating consequences: it leads to physical injuries, psychological stress, decreased motivation to work, and staff absenteeism at a time when they are most needed. Disturbing incidents, including fatal attacks on doctors (e.g., the murder of a doctor in Krakow in 2025), have highlighted the scale of the problem and prompted calls from the medical community for decisive action. The problem of aggression against medical professionals is global in nature and is growing in many countries. The data is alarming: in Poland in 2023, almost 40% of doctors working in hospital emergency departments (EDs) and admission rooms reported at least one case of aggression by a patient or their relatives.

Western European countries are also struggling with a similar intensity of the phenomenon. According to the WHO, up to 38% of healthcare workers worldwide experience physical violence during their careers. In the United States, as many as 72.8% of all serious cases of workplace violence (requiring time off or a change of position) occur in the health and social care sector. These statistics prove that aggression toward doctors is a common and growing phenomenon in many healthcare systems, regardless of their level of funding.

In order to prevent acts of violence against medical staff, hospitals are implementing formal procedures for responding to incidents of aggression. Staff have clear instructions on when to call security or the police, how to record the incident, and how to secure evidence. Internal incident reporting systems are being created (e.g., aggression reporting forms for doctors in France or the Staff Observation Aggression Scale - Revised (SOAS-R) system for describing acts of aggression used in hospitals in Germany). In addition, some facilities keep warning lists or notes in patient records that flag individuals who have been aggressive in the past, allowing staff to prepare for potential risks on subsequent visits. Such solutions are in place, for example, in British primary care as part of the *Special Allocation Scheme*, a program for referring aggressive patients to special facilities. The overriding principle is a zero-tolerance policy for violence – in many countries, doctors and management are encouraged to report any serious threat or attack to the police, rather than treating them as an "occupational hazard." As a result, more cases are referred to law enforcement agencies, which increases the chance of bringing aggressors to justice and acts as a deterrent.

In many European countries, communication and de-escalation training is standard; for example, the British NHS is developing training programs for healthcare workers in conflict resolution, and a comprehensive approach combining skills training with other safety measures is considered to be the most effective.

It is an indisputable fact that the well-being of medical professionals is a key element of workplace safety in the face of increasing patient aggression. Constant tension, fear of attack, and repeated incidents of violence lead to burnout, sleep disorders, anxiety, and depression in many healthcare workers. A tired and stressed doctor or nurse is more prone to mistakes and may find it more difficult to deal with conflict situations. Therefore, emotional and organizational support, such as access to a psychologist, the opportunity to discuss difficult situations within the team, flexible schedules, and time for regeneration, is just as important as training and technical safeguards. It is becoming standard practice to create well-being programs in healthcare that include both post-violence support and preventive measures (anti-stress workshops, mindfulness, and promotion of physical activity). Caring for the well-being of doctors is the foundation for building a safe, stable, and more empathetic work environment.

It seems that the best results are achieved by those healthcare systems that have implemented multifaceted protection programs: combining training (improving staff skills and reducing stress) with investments in security infrastructure (monitoring, alarms, and physical protection), appropriate procedures (clear guidelines and incident reporting), and legal and institutional support (severe penalties, zero-tolerance policies, and psychological assistance for victims). Western European countries and the United States have numerous examples of effective solutions – from special police units in hospitals to modern wearable technologies – that have improved the comfort and safety of doctors' work, reducing both the number of violent incidents and their severity. The Polish healthcare system, which faces similar problems, can draw inspiration from these experiences. Although not all solutions can be easily transferred (e.g., due to costs), many of them – such as de-escalation training, the installation of cameras, and alarm systems – are feasible to implement and can significantly improve the situation. Providing doctors with safe working conditions is essential so that they can fully focus on their mission of saving patients' lives without fear for their own health and safety.

The publication presents research findings, procedures, and guidelines for increasing the safety of medical personnel in response to current statistics on aggression against medical professionals, as well as an analysis of the effectiveness of these measures – that is, their impact on doctors' sense of safety. This is an important publication that can help shape a sustainable work environment, one in which medical personnel can work safely and develop their professional passions while helping others.

Professor Agata Ptak-Belowska, MD
V-Dean for English-Language Education
Second-degree specialist in internal medicine with many years of experience in hospital wards, emergency departments, and primary care. Currently, she focuses on research and teaching, heading the English-speaking Faculty of Medicine and Dentistry at the Jagiellonian University Medical College in Krakow, Poland.

Acknowledgment

We would like to express our sincere thanks to all the hospital directors who agreed to take part in the study, the results of which are presented in this book.

Reviewers of the edition:
Professor Iwona Kowalska-Bobko
Professor Ewa Wierzbicka-Mazur

Introduction

This book is the result of both extensive academic research and our professional experience. The cognitive exploration of the topic has reinforced our belief that one of the key challenges in building a sustainable healthcare workplace is management action to reduce the phenomenon of aggressive patient behavior. To successfully effect such action, it first requires examining the problem from a variety of angles. The basis, of course, is knowledge of how healthcare systems work. It enables effective identification of areas and situations with a high likelihood of aggressive patient behavior. Second of all, psychological knowledge is vital to understand the most common mechanisms of aggressive behavior. Last but not least, it remains an imperative to be ethically attentive, always remembering that even the most aggressive patient is first and foremost a human being in need of help. Working in an interdisciplinary team, we have tried to take these different points of view into account.

The book begins with a chapter on the particularities of working in the health sector. We characterize the challenges faced by workers in this sector. We also present the debates on the nature of medical services, which revolve around the question: "Who is the recipient of medical services – the patient or the client?"

The second chapter deals with occupational safety issues in the broadest sense. Starting with the definitional issues of the concept, we identify the most common risks and methods of managing them. We treat these activities as prerequisites for creating a sustainable healthcare workplace.

Chapter 3 focuses on hazards in the healthcare work environment. We discuss the physical, biological, chemical, and psychosocial factors that affect the safety of healthcare workers. We pay particular attention to the latter group of hazards, including the results of research showing that the likelihood of making a medical error is twice as high after 12 hours of work than after 8 hours.

Chapter 4 is entitled "The Different Faces of Violence and Aggression in Healthcare." The starting point here is a review of theories of aggression and the characteristics of its different forms, such as passive and active aggression, direct and indirect, hostile (reactive) and instrumental (proactive). Three categories of aggression are also distinguished, such as (1) manifest hostility, (2) obstructing and preventing the performance of tasks, and (3) overt aggression, which has the nature of physical aggression. This chapter also characterizes the phenomenon of so-called cyber-aggression, which is increasingly affecting medical personnel, especially doctors.

The fifth chapter of the book is devoted to presenting the methodology of the research conducted and analyzing the answers given by the respondents. The results of the research were shocking – almost 95% of medical staff experienced aggressive behavior from patients. This state of affairs has very serious consequences – it leads to disorganization of the work of the medical institution (73% of responses), increases the risk of medical errors (61%), and causes a decrease in work efficiency (54.5%). The decrease

DOI: 10.1201/9781003716631-1

in motivation of medical staff also proved to be a major problem – over 50% of respondents felt dissatisfied with their work and almost 30% felt that their dignity and professional pride had been taken away. In this chapter, we also discussed the phenomenon of so-called organizational silence, which is a kind of breeding ground for various types of pathological behavior in the workplace.

Chapter 6 presents conclusions and practical recommendations on management actions that should be taken to reduce the phenomenon of aggressive patient behavior. The feasibility of these actions is supported by the examples of good practice described in this chapter. Based on analysis of many of these practices, research and our own professional experience, we present a process for implementing the necessary changes to support the creation of a sustainable healthcare workplace.

Counteracting aggressive patient behavior is an issue whose social importance cannot be overestimated. Highly qualified medical personnel should remain in constant readiness to carry out the task for which they were appointed – saving human life and health. Meanwhile, doctors, nurses, and paramedics devote a huge part of their time and energy to struggling with aggressive behavior from the people they want to help. To change this, lofty slogans or one-time actions are not enough. Systematic management actions based on interdisciplinary knowledge are necessary. We hope that reading this book will be helpful in taking such actions.

Ewa Beck-Krala, Elżbieta Tarczoń,
Małgorzata Sidor-Rządkowska,
and Jerzy Rosiński

1 Management Challenges in the Healthcare Sector

Civilizational development brings many changes. Alongside the positive aspects of development, there are also negative ones, including new threats in the workplace. In the healthcare sector, in particular, employees face a greater number of threats, against which they are often helpless. Identifying the most important risks will help to effectively counteract them in the future and manage employee safety, supporting the well-being of both healthcare workers and patients.

INCREASING DEMANDS IN HEALTHCARE WORK

Creating a supportive work environment and managing occupational safety is currently one of the most important areas of EU social policy [EU-OSHA 2021[1]]. Especially after the SARS-CoV-2 pandemic, many employers have begun to prioritize this issue by eliminating emerging risks and creating a culture of safety in the workplace. Employers' attention to this area can have a positive impact on employee engagement, productivity, and health, and ultimately translate into higher financial results for the organization. Dynamic changes in the environment and unexpected crises, which have been occurring with great intensity recently, generate additional risks and pose a huge challenge for employers and employees themselves. In turn, low levels of safety in the workplace lead to reduced productivity and quality of work, increased employee absenteeism, and higher organizational costs [EU-OSHA 2021[2]]. A large scale of threats to the health and life of employees is observed especially in the healthcare sector, which has been recognized as one of the most dangerous [Wiskow et al. 2010, p. 13]. The working environment in this sector is unpredictable due to the scope and type of tasks performed [Sidor-Rządkowska 2018, p. 131]. In addition to physical, chemical, and biological hazards, there are an increasing number of psychosocial hazards. These risks are a consequence of rapid changes and crises, as well as emotional challenges resulting from both the increased volume and difficulty of work in customer-oriented sectors. The most common psychosocial burden seems to be so-called difficult customers [EU-OSHA Report 2023[3]], which in healthcare can manifest itself, among other things, in aggressive behavior by patients and their family members, to which medical staff are exposed.

The upward trend toward more psychosocial and emotional challenges at work does not mean that the "classic" (existing) risks of accidents in the workplace have disappeared. They occur in parallel, so to speak, increasing the difficulty of work and strongly affecting the quality of work. A fairly constant percentage of workers continue to report exposure to physical, chemical, and biological factors; depending on the occupation and sector, between 15% and 30% of workers are exposed to such risks [EWCS[4]]. Over the past 15 years, there has been no or very little decrease in this risk.

DOI: 10.1201/9781003716631-2

A very important factor affecting health and safety at work is the length of daily or weekly working time and shift work, especially night shifts. All these factors are present in the work of healthcare workers, which makes this professional group particularly vulnerable to life- and health-threatening dangers but also to high emotional costs.

In healthcare facilities, patient safety is a priority. Unfortunately, the pursuit of this goal sometimes causes harm to the safety and health of the employees themselves [Mikos 2017; Witczak, Rypicz 2020]. Creating safe and hygienic working conditions in the healthcare sector mainly boils down to implementing the provisions of applicable laws and regulations. However, there is a lack of comprehensive measures to identify risks, especially psychosocial risks associated with aggressive patient behavior.

This risk is particularly high in hospital emergency departments (EDs). Violence in hospital EDs around the world has been escalating for many years [Rabin et al. 2025]. A stay in the ED is an extremely stressful situation for patients and is usually associated with intentional or unintentional aggressive behavior on the part of patients or their relatives [Rabin et al. 2025]. Due to the frequency of such incidents, they have been called an inevitable occupational hazard [Rabin et al. 2025]. Currently, every ED employee is aware that their work has a dark side: violence against ED staff [Ramacciati et al. 2019]. According to the National Institute for Occupational Safety and Health, "Acts of violence (including physical assaults and threats of assault) directed against persons at work or on duty" are considered workplace violence (WPV) [Ramacciati et al. 2019]. This phenomenon in healthcare facilities, especially in EDs, is so widespread around the world that it has reached epidemic proportions [Ramacciati et al. 2019; Tarczoń et al. 2025], which, unfortunately, is growing in intensity [Rabin et al. 2025]. As a result, many international organizations, such as the International Labor Office, the International Council of Nurses, the World Health Organization (WHO), and Public Services International, are publicizing this problem in an effort to raise awareness among various social entities in this regard.

Hence, this book discusses the risks and consequences associated with aggressive patient behavior experienced by medical staff in hospital EDs. This issue has also been addressed due to the lack of legal regulations in many countries (including Poland) on how to deal with patients whose behavior is socially unacceptable.[5]

Less attention is paid to the safety of medical staff than to the safety of patients [Witczak, Rypicz 2020, p. 11]. From the patients' perspective, this has a positive impact on the image of the medical facility, but it is not conducive to hiring and retaining the best employees. The increasing number of cases of aggressive patient behavior has a negative impact not only on the safety, well-being, and job satisfaction of medical staff, but also on the quality of medical services and care provided. Therefore, the book focuses on analyzing the risks and consequences associated with aggressive patient behavior in hospital EDs.

CHALLENGES FACING THE HEALTHCARE SECTOR

The healthcare sector is a strategic sector of the economy that aims to provide services that meet the healthcare needs of the population and reduce the incidence of chronic diseases. It is of fundamental importance to both society and the economy in every country in the world. This sector usually generates a large number of jobs and contributes to social and economic development. Currently, around 10% of workers in the European Union (EU) are employed in the healthcare and social protection sector [EC 2011].

Despite this, the sector is experiencing an increasingly acute shortage of workers, which has and will continue to have enormous consequences for society. The growing demand for medical services caused by longer life expectancy and innovations in medicine is further increasing the need for medical services.

The number of doctors in EU countries has risen to 1.83 million, but this is still insufficient to meet the needs of the population. According to estimates, in 2022, there was a shortage of 1.2 million doctors, nurses, and midwives in EU countries. In all EU countries, nurses constitute the largest group of healthcare workers. In 2022, there were 8.4 nurses per 1,000 inhabitants on average, which is a slight increase compared to previous years [*Health at a Glance: Europe 2024, State of Health in the EU Cycle*, p. 194]. The period of professional activity of doctors and nursing staff is also lengthening, and the percentage of retired doctors is increasing. A process of feminization of the professional group of doctors and nursing staff has also been observed. In 2022, more than 53% of doctors in EU countries were women, who worked about 10% less than male doctors. On average, women worked 39.4 hours per week compared to 43.0 hours for men [*Health at a Glance: Europe 2024, State of Health in the EU Cycle*, p. 188]. The vast majority of nursing staff are also women, who account for over 85% of those working in this profession. Nursing staff significantly outnumber doctors in most EU countries. Currently, the number of doctors in EU Member States per 1,000 inhabitants ranges from 2.4 to 5.4, while the number of nursing staff per 1,000 inhabitants ranges from 4.4 to 15.4 [*A European Health Union. A Blueprint for Generations* 2023, p. 13]. An important challenge for management is to attract young people to the medical professions. The average growth rate of new nursing graduates in the EU is only 0.5% per year [*Health at a Glance: Europe 2024: State of Health in the EU Cycle*].

Emergency medicine plays an important role in the healthcare system. Medical services in the ED account for 10% of all visits outside the hospital and 26.1% of all outpatient visits to the hospital. During the year, approximately 28% of the population uses the assistance of medical staff in EDs, and this number is growing by 2%–3% annually. On average, 304 people per 1,000 inhabitants use the medical services provided in EDs [*European Emergency Medicine in Numbers. Emergency Medicine. Epidemiology Series*]. The Commission and the Organization for Economic Cooperation and Development (OECD) have published a report highlighting the urgent need for action to address the shortage of healthcare workers. Action is also needed to provide support for workers. This action is required at several different levels, particularly in terms of greater investment, better working conditions, and greater training opportunities [*Health at a Glance: Europe 2024: State of Health in the EU Cycle*].

Over the past decades, the healthcare sector has supported the protection and restoration of public health. Currently, this sector is undergoing many transformations that pose both challenges and opportunities for its development. The most visible changes are demographic shifts and population growth, as well as rising healthcare costs worldwide. Increased life expectancy has led to a greater burden of chronic diseases and other health problems on citizens, which is why the demand for medical services is growing year on year, while access to medical services is becoming increasingly limited. There is an imbalance between the demand for medical services and the ability to provide them. The main challenges facing the healthcare system are demographic issues, i.e., an aging population and the development of the most expensive form of medical care, which is hospitalization.

The problem of an aging population is also evident among healthcare workers. There are serious concerns about the shortage of nurses and midwives. It is estimated that by 2033, there will be a shortage of 169,000 nurses and midwives [de Jong et al. 2014, p. 15].

Many countries have also seen significant changes in the labor market and its environment. There has been a period of free migration of citizens, flexible forms of employment, a decline in unemployment, and a shortage of workers. According to the European Agency for Safety and Health at Work (EU-OSHA), over the past 30 years, the number of migrant healthcare workers has increased by more than 5% per year [de Jong et al. 2014, p. 15]. As a result of worker migration and an aging population, there is a need to "manage" workers appropriately and to seek new methods of recruiting candidates for jobs in the healthcare sector as the healthcare sector is one of those experiencing the greatest shortage of workers.

A worrying phenomenon in some countries is the migration of doctors after graduation, specialists with several years of experience, and highly qualified nursing staff. The migration of healthcare workers may further exacerbate the existing labor shortage and lead to a greater workload for the remaining employees. Due to the existing shortage of workers, the solution is to hire immigrants. A positive consequence of the migration of medical workers is the opportunity for employees to improve their skills. Cooperation between healthcare workers allows them to train each other in new medical procedures [de Jong et al. 2014, pp. 15–18].

In addition to the migration of medical workers, another threat to the healthcare sector is the economic migration of citizens and changes in the family model in many countries. Women entering the workforce and migration have meant that older people can no longer count on the help and support of family members [ILO 2019; Institute for Labor Market Analysis 2020]. This increases the demand for healthcare and contributes to the growing need for care provided by healthcare workers. The healthcare sector cannot change certain working conditions, such as night work, weekends, and holidays; other incentives will be needed to attract young people to work in this sector [Wiskow et al. 2010, p. 2]. Therefore, the further development of the healthcare sector is becoming inevitable. The problems facing healthcare indicate the need for change, if only because of its role and importance, which is likely to increase.

Emergency medical services, including hospital EDs, which operate within the infrastructure of hospitals, play a key role in the healthcare system. They are designed to save human life and health in emergency situations. It is a requirement to provide 24-hour and immediate access to diagnostic tests performed in a medical diagnostic laboratory, computed tomography, and endoscopic examinations, including gastroscopy.

THE SPECIFIC NATURE OF WORK IN HEALTHCARE

Working in healthcare involves constant contact with patients who are experiencing pain, suffering, and fear for their lives and health. Working in healthcare requires maintaining contact with the family members of sick people, who are often nervous and frustrated. Therefore, it is difficult, full of unpredictable situations and traumatic experiences related to saving human life and health [Sidor-Rządkowska 2018, pp. 131–132]. An integral part of this work is time pressure and stress related to responsibility for the patient's health, entrusted medical equipment, medicines, medical supplies, and medical procedures performed. An important aspect of the job is the necessary continuous development of the professional competence of medical staff, which is inherent in the specific nature of this work. The high demands of the job mean that medical staff should be physically and mentally resilient, assertive, and committed. Given the high level of mental and physical strain (stress), responsibility, and new threats, it is difficult to maintain a healthy work-life balance.

The work of medical staff is varied and differs from one unit to another. In hospitals, it is common to work in shifts and to be on call. This means that there are few staff members in the wards and other organizational units of the hospital. Night work requires staff to be ready to act, and care for more patients than during the morning (teamwork). In addition, night work disrupts the circadian rhythm and requires increased attention when performing care and treatment tasks.

Work in hospital EDs is particularly stressful, as this is where patients first come into contact with medical staff. As the name suggests, the "emergency" department deals with patients in immediate danger to their life and health. In every ED, upon arrival and admission to the ward, patients are triaged according to a five-level **triage system**. The segregation is carried out by qualified medical personnel: nursing staff or paramedics, based on a preliminary assessment of vital signs (heart rate, breathing, level of consciousness, and ability to walk). Thanks to this simple method, staff can provide assistance to patients whose health is most at risk. In this case, providing assistance involves the provision of medical services of a special nature.

RECIPIENTS OF MEDICAL SERVICES

There is some controversy surrounding the specific nature of medical services, which is also connected with the recipients of these services. There is an ongoing discussion whether the recipient of medical services is a patient or already a client. Medical services are a specific type of service, different from health services. Health services are part of medical services and are defined as activities aimed at preserving, saving, restoring, or improving health, as well as other activities resulting from the treatment process [Wiśniewska 2016, p. 21]. According to the definition, health services are directly related to treatment processes and activities. Medical services, on the other hand, include elements of health services extended by other factors that allow the patient to achieve satisfaction [Wiśniewska 2016, p. 22].

Features of Medical Services

One of the characteristics of broadly understood medical services is uncertainty in the sphere of demand. The demand for this type of service is associated with an absolute necessity and the very broad range of services provided by medical personnel. The basic feature of medical services is the professionalism and adequate competence of the staff. The provision of medical services requires specialist knowledge, experience, skills, and continuous improvement. Medical services are unique in nature, which means that medical personnel perform their duties while making decisions about the health and lives of other people. A feature of medical services is their individual nature and adaptation to the needs of specific patients, their conditions, and their state of health. Another feature is the lack of substitutes. Medical services cannot be replaced by others, and their provision requires the use of various medical techniques and technologies, appropriate diagnostic methods, and IT solutions. Other features of medical services include psychological complexity and limited quality control. In addition, they are associated with high risk, uncertainty, anxiety, and the patient's condition, and depend on the patient's behavior and expectations [Krot 2008, pp. 15–16; Wiśniewska 2016, p. 25]. An important feature of medical services is their value, which is created as a result of interaction between the service recipient and the service provider, and influences the treatment process and patient satisfaction. Another area related to the provision of medical services is the impact of external effects. It is difficult to prove the influence of the external environment or patient behavior on the occurrence of specific diseases and complications. Another feature of medical services is their inseparability. This means that a medical service is provided by medical personnel in the presence of the patient, and its delivery and consumption take place at the same time and place. Medical services are variable because even the same service (medical procedure) is different each time, despite being provided in the same medical facility and by the same medical staff. A specific feature of medical services is the difficulty of assessing them (purchase uncertainty). This feature is related to the inability to adequately assess the competence of a doctor. An inherent feature of medical services is risk, which in this case means uncertainty about the effects of treatment [Krot 2008, pp. 15–16; Wiśniewska 2016, p. 25; Sidor-Rządkowska 2018, p. 132]. What is more, one of the problems related to medical services is liability. Medical personnel play a key role in the process of providing medical services – their behavior toward the patient, their competence, kindness, willingness to help, and support in dealing with non-medical problems that usually accompany illness [Sidor-Rządkowska 2018, p. 131].

Patient or Customer

The above characteristics of medical services make the terminology used to describe people in contact with healthcare problematic. In healthcare entities, the terms "patient" and "customer" are often used, and it is not easy to distinguish between them. Commonly, the word "patient" is reserved for a person using medical services, while "client" or "customer" means a person using professional services offered by business entities [*10 Differences...* 2024]. The term "patient" means a person who has or shows patience (from Latin *patiens*) and a person who is waiting for or receiving medical care [Deber et al. 2005, p. 346]. The term "client," on the other hand, comes from the Latin word for dependence and is defined as a person under the protection of another person or using professional advice or services [Deber et al. 2005, p. 347]. The first definition of the term "patient" was proposed by the WHO in the Declaration of Patient Rights in Amsterdam in 1994. According to its provisions, a patient is a person who uses healthcare services, regardless of whether they are healthy or sick [Czerwińska 2015, p. 343]. The basic requirement of patients is professional medical care consisting in the diagnosis of their diseases, maintenance of their general health, or its improvement. A customer, on the other hand, focuses on achieving a goal, obtaining assistance with a specific service or product. This focus on maintaining or improving health distinguishes a patient from a customer.

In the healthcare sector, there are strong relationships between medical staff and patients, manifested in subordination, trust, mutual respect, empathy, and care that encompasses the personal aspects of the patient's well-being and safety. In the context of dynamic changes in the healthcare sector, the concept of the "customer approach to healthcare" is becoming increasingly prevalent, emphasizing individual needs and preferences and active participation in the decision-making process [*10 Differences...* 2024]. In turn, the relationship between the customer and the service provider depends on the quality of the service or product, price, customer satisfaction, and industry. In the healthcare sector, this relationship is based on mutual trust, which is a prerequisite for effective medical care. The decision-making process itself is one-sided due to the patient's lack of medical knowledge. In the case of the customer, the decision-making process is based on cooperation – the customer determines the scope of work and presents their expectations. In the healthcare sector, the relationship between the patient and the healthcare provider is private. Respect for patient confidentiality and medical records is a priority that applies to all entities providing medical services.

In the medical services sector, the relationship between the patient and the healthcare provider tends to be long term. Patients use the services of the same primary care physician or pediatrician for many years. The same applies to specialist services. In the context of healthcare, an important aspect is the rights of the patient, which ensure respect, dignity, access to information about their health, risk factors, treatment options, and the confidentiality of this information, as well as the right to refuse certain therapies. The most important is the patient's right to give informed consent to treatment, which protects them from unwanted interventions. Customers, on the other hand, expect good service, transparency in contracts, fair and equitable treatment, and the proper handling of any complaints within a reasonable time.

Patients are people who are suffering, in need, and expecting professional medical help, which exposes them to commercial treatment. This is a worrying phenomenon as it suggests that the provision of medical services is commercial in nature. Emergency medical teams use the term "patient" to refer to people requiring medical assistance. In this context, patients require care, while customers have a choice of available services and products. Rehabilitation specialists, on the other hand, prefer the term "client," as decision-making in this area is mutual. The professional group of doctors and nursing staff is heterogeneous, but most of them use the term "patient" [Ratnapalan 2009, p. 472]. The debate between "client" and "patient" continues in the literature on the subject. The term "client" has gained a certain dominance in social media and therapeutic relationships over "patient." This term was introduced by Carl Rogers, an American psychologist and psychotherapist, the creator of "client-centered psychotherapy" [Costa et al. 2019, p. 1]. The author claimed that a person participating in psychotherapy is a partner of the psychotherapist. In this case, the term "client" rather than "patient" is used to emphasize the partnership and at the same time prevent the stigmatization of the sick person. The use of the term "client" makes the suffering person a buyer, service recipient, petitioner, or customer.

Another issue concerning patients and clients is emotional bonding. Patients with health problems expect compassion, understanding, and sensitivity from medical staff. When interacting with patients, medical staff usually share their concerns about the correct diagnosis and effectiveness of treatment, as well as sharing their joy when the treatment is successful. The bonds between clients and service providers are less emotional, although a certain level of camaraderie may develop over time, but this does not translate into personal emotions. In the customer-supplier relationship, the parties focus primarily on the quality of services.

The healthcare system is an area regulated not only by legislation in individual countries but also by professional medical associations. In the case of customers, however, each industry adapts its standards and regulations to the requirements and expectations of customers.

Although the terms "patient" and "customer" are linked by a similar relationship, their dynamics, responsibilities, and expectations differ significantly. A patient seeks medical services, while a customer seeks a service or product. In the first case, the relationship may last a lifetime, while in the second case, it may only last for a specific period of time. It all depends on the nature of the relationship, the needs and expectations of the individual, and industry regulations. The American Psychological Association recommends using the term "patient" when describing or discussing a person or population of people who are receiving healthcare services or who are participating in health-related research [APA Resolution 2018].

In summary, medical services are a specific type of service and cannot be reduced to commercial services without compromising the quality of the services offered and patient safety. Patients often do not have the opportunity to choose a medical service due to systemic limitations; they are suffering, in need of help, and therefore vulnerable to exploitation in a purely commercial model. Furthermore, they are unable to clearly determine whether the optimal treatment has been implemented. Due to this asymmetry of relations and information between the patient and the healthcare system, it cannot be reduced to the use of services. Previous research results also indicate that most patients prefer to be called patients rather than customers [Costa et al. 2019, p. 4].

In the context of the issue at hand, treating patients as customers may additionally generate more undesirable behaviors. The use of commercial metaphors gives permission for demanding attitudes toward medical staff within the framework of a paid service.

THE INTANGIBLE WORK ENVIRONMENT OF MEDICAL STAFF

Each healthcare facility has a unique atmosphere, culture, and environment in which there are factors that affect the health and lives of all employees, regardless of their roles.

One of the key factors in the intangible work environment of healthcare workers is working time. Working time is not only the time during which an employee works but also the time during which an employee is ready to perform work. In healthcare facilities intended for people whose health condition requires 24-hour healthcare services, medical professionals with higher education are required to perform medical duties, the time of which is included in their working time.

Another important factor in the intangible work environment in healthcare is the quality of human resources management processes, such as employment planning, recruitment and selection, remuneration, employee evaluations, training and development, and employee turnover [Armstrong 2007, p. 331; Armstrong, Taylor 2016, p. 269]. Recruitment is a key factor in the functioning of a medical facility. Medical facilities are looking for employees with very high competences and the right qualifications.

Another function performed as part of human management in an organization is employee development, including training. People working in the medical professions are required to undergo continuous professional development. Due to the enormous progress in the fields of medicine, the knowledge gained during academic studies is insufficient. The process of professional development, especially in the area of new achievements and therapies as well as treatment standards, is a prerequisite for the quality of medical services provided. Unfortunately, in some countries, the participation of healthcare workers in training is considered a cost rather than an investment in employees who will perform their duties more professionally in the future and pass on the knowledge they have acquired to their colleagues, which will translate into higher quality services [Stańczyk 2012, p. 126]. Training offered to healthcare workers is usually focused on developing so-called "hard skills" (specialized skills). However, there is a lack of training in soft skills such as communication, stress management, and decision-making. In turn, employee evaluation activities in healthcare facilities are schematic in nature. They result from the requirements of the quality management system or accreditation [Buchelt-Nawara 2006, p. 661].

The shortage of healthcare workers has contributed to the development of research into the factors determining people leaving the medical profession. Dissatisfaction with salary levels is particularly evident among nursing staff. In several European countries, 58%–90% of nursing staff have expressed such dissatisfaction [Wiskow et al. 2010, p. 1]. Salary is not the only factor determining leaving work. The decisive reasons for nursing staff leaving the profession are lack of respect, dissatisfaction with working conditions, and the low status of the profession. Sixty-one percent of the nursing staff surveyed complained about the low status of the profession [Gawęda, Śnieżek, Serzysko 2018, p. 271]. A study of paramedics yielded similar results, with 71% of paramedics considering the remuneration system to be the most mentally stressful aspect of their work and 57% pointing to limited opportunities for promotion and professional development [Żurowska-Wolak et al. 2015, p. 46].

An important factor in the intangible work environment in the healthcare sector is interpersonal relationships, which affect the level of satisfaction and contentment with professional work [Becker-Pestka et al. 2017, p. 19]. Informal relationships are extremely important in healthcare facilities, as most of the work is team-based. It requires cooperation between doctors, nursing staff, paramedics, and other employees not involved in the treatment process (e.g., registration staff). Relationships between medical staff and patients and their families are very important. This contact includes interviews, providing information, performing specific medical procedures, but also informing about further treatment or test results. These conversations are often accompanied by difficult emotions and even conflict, which can be difficult to manage and which has a strong impact on medical staff and their work.

SUMMARY

The healthcare sector plays an important role in every society. It performs tasks related to providing services to patients that promote the preservation and restoration of their health. Maintaining health and restoring life is achieved not only through the use of modern medical advances but above all thanks to the high competence of medical staff (specialists) from various fields of medicine. Their work is extremely responsible and demanding, and therefore very stressful. It is also related to the diversity of medical services provided, for which demand is growing as people become more aware and their life expectancy increases. Changing patient preferences and behaviors indicate a need for more medical services and greater involvement on the part of medical staff. Unfortunately, due to the insufficient number of medical personnel, including specialist doctors, the provision and availability of many medical services is limited. This raises concerns about the quality and availability of medical services in the future, especially in view of the emerging new threats and stressors in the working environment of medical personnel. In addition, the low social status and age structure of key representatives of the healthcare sector suggest that the problem of access to medical services for the public will grow. This is especially true in view of the increasing challenges in this work caused by psychophysical requirements and threats related to the intangible work environment.

NOTES

1 *EU*-OSHA 2021 (https://osha.europa.eu/pl/themes/good-osh-is-good-for-business 17.03.2022).
2 *EU*-OSHA 2021 (https://osha.europa.eu/pl/themes/good-osh-is-good-for-business 17.03.2022).
3 EU-OSHA Report 2023, *Occupational safety and health in Europe: Status and trends 2023* (https://osha.europa.eu/sites/default/files/OSH_in_Europe_state_trends_report_2023_en.pdf 13.12.2023).
4 *European Working Conditions Survey* (https://www.eurofound.europa.eu/en/topic/living-and-working-europe 27.11.2023).
5 An exception is the behavior of persons with mental disorders who commit an attack against their own life or health or the life or health of another person, public safety, or violently destroy or damage objects in their surroundings.

2 Work Safety as an Element of a Sustainable Healthcare Workplace

A sense of safety at work is one of the basic needs of every employee. Work safety is understood not only in terms of permanent employment that provides income, but also as protection against hazards occurring in the course of work. Work safety is one of the main aspects of creating a sustainable work environment, especially where it is at risk, as in the healthcare sector.

SUSTAINABLE WORK ENVIRONMENT

The prerequisite for building a sustainable work environment is caring for the well-being of the people employed in the organization. This multidimensional concept applies to various areas of life. It can be considered at the individual, organizational, and societal levels. Attempts by various authors to define this concept have not led to a consensus. However, the essence of the issue is well captured by the World Health Organization's definition, according to which it is a state of being in which a person *understands their capabilities, copes with life stress, works productively, and contributes to their community* [Juchnowicz, Kinowska 2022, p. 35].

The concept of human well-being *was* initially developed within the framework of positive psychology, a field pioneered by Martin Seligman, president of the American Psychological Association [Tabor-Błażewicz 2022, p. 48], who stated that psychology should not focus only on states of deficiency and disorders, but also examine the conditions for achieving human well-being. This state was defined, among other things, using the PERMA model, which stands for:

P positive emotions
E engagement
R positive relationships
M meaning
A accomplishment

Well-being can be viewed in a hedonistic and eudaimonistic way. The former (not coincidentally associated with the philosophical trend known as hedonism) means a state of well-being and satisfaction. Well-being (derived from Aristotle's eudaimonia), on the other hand, refers to the realization of human potential. The key values here are: a sense of meaning in one's actions, personal development, and positive relationships with other people. In simple terms, hedonistic well-being means a pleasant life, while eudaimonistic well-being means a meaningful life.

DOI: 10.1201/9781003716631-3

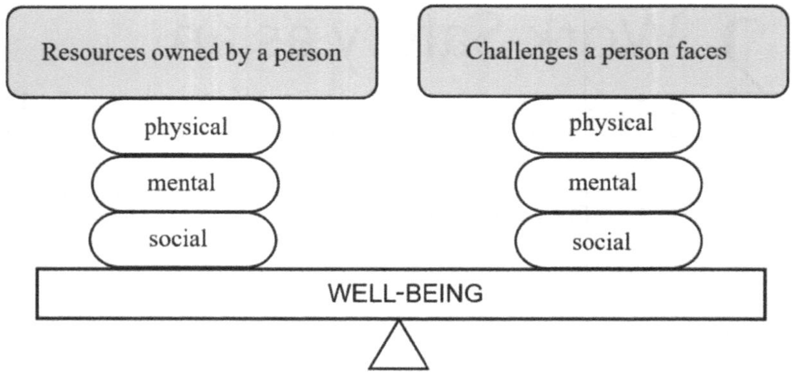

FIGURE 2.1 Wellbeing as a state of balance between resources and challenges

R. Dodge, A.P. Daly, J. Huyton, and L.D. Sanders define well-being as "a state of balance between an individual's resources and the challenges they face." Both resources and challenges can be physical, mental, or social in nature. This principle is illustrated in Figure 2.1.

Well-being in an organizational context is a new and still not fully recognized phenomenon. It focuses mainly on shaping working conditions aimed at maintaining the health and improving the well-being of employees, as well as having a positive impact on their professional development, sense of security, and building relationships between them.

Well-being is a state of balance, so a sustainable work environment can be understood as one where a healthy balance is created between the demands of the job and the resources that employees have at their disposal to perform their tasks. This means creating appropriate working conditions, i.e., ensuring physical and mental comfort in the workplace, which has a strong impact on employee health. Physical comfort concerns aspects such as broadly understood material working conditions, which in the case of medical staff means appropriate ergonomic and safe workstations, which is related to, for example, proper lighting, access to efficient medical equipment, protective measures, ensuring adequate breaks and rest during long shifts, or even the possibility of eating healthy meals during work. Equally important is psychological comfort, which is related to stress resulting from the fast pace of work and the difficulty of medical procedures, the quality of relationships between colleagues and patients, teamwork and the prevailing work atmosphere, or a sense of agency and job satisfaction. However, healthy working conditions are primarily about safety in the workplace, i.e., the elimination of threats to physical and psychological health and safety.

A lack of safety during the performance of daily professional duties means a drastic violation of the aforementioned balance. Representatives of international associations dealing with employment conditions emphasize that the elimination of hazards should be the starting point for building a sustainable work environment. For example, according to the European Working Conditions Observatory, such an environment means "safe, healthy, and productive work in a well-managed organization, performed by competent employees who perceive their work as meaningful and satisfying, and whose work is an element that supports their lives" [EWCO].

According to representatives of the International Labor Organization (ILO), well-being at work is a multifaceted concept, covering both the quality and safety of performing professional duties, as well as the way of thinking about one's work, the atmosphere at work, the way tasks are performed. etc. ILO representatives emphasize that employee well-being is a key factor determining the long-term effectiveness of an organization [Kulig-Moskwa 2018, p. 14].

The concept of well-being in an organization is as difficult to define as the concept of well-being in general. One of the basic dilemmas is whether employee well-being should be considered from a hedonistic or eudaimonistic perspective. As Magdalena Jaworek (2021, p. 38) points out, in recent years there has been a tendency to use the latter approach. This is related to the development of research and theoretical analyses on issues such as professional self-fulfillment, commitment to work, and a sense of meaning in work.

An employee's well-being can be classified into one of three basic categories. The first is negative, characterized by experiences such as excessive stress, workaholism, burnout, and mobbing. The second is a neutral "zero state" characterized by the absence of negative experiences. The third category means well-being – the individual experiences positive emotions and a sense of commitment to work. These categories are shown in Figure 2.2.

Many authors (including Czerw 2017; Staszkiewicz, Klimkiewicz, Beck-Krala 2023) point out that the approach presented is simplified. It is difficult to disagree with the statement that

> (…) the concept of treating ill-health/illness/ill-being and *well-being* as the extremes of a bipolar continuum is a matter of debate. Research on positive and negative affect has shown that they are independent of each other and that the absence of symptoms of illness is not the same as well-being.
>
> **[Staszkiewicz, Klimkiewicz, Beck-Krala 2023, p. 166]**

However, it is worth remembering that the presented model can play a useful role in reminding us of the need to take preventive measures to prevent the emergence of negative phenomena, the extreme manifestation of which is a lack of security in the work environment. The lack of safe working conditions, both physical and psychosocial, ensuring a balance between work and leisure time, guaranteeing fair wages

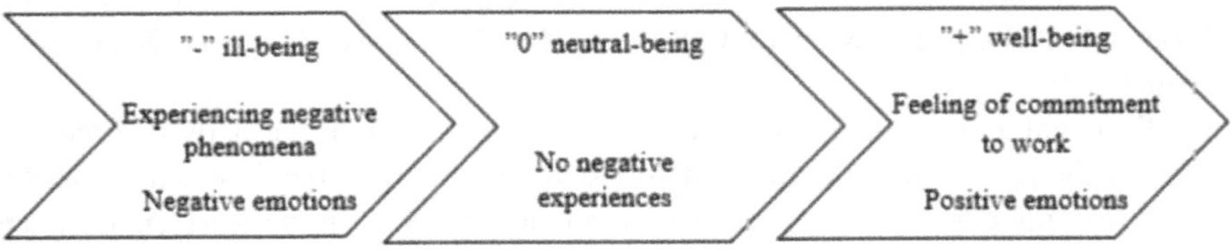

FIGURE 2.2 Three categories of employee well-being

and protecting the health of the employee, means that the employee does not achieve the psychological benefits of their work.

THE CONCEPT OF OCCUPATIONAL SAFETY

In the literature on the subject, the concept of safety is considered to be "a state in accordance with legal regulations, the purpose of which is to ensure safety, or a state in which hazards are kept under control" [Liwo 2013, p. 18]. This means that safety is "a state in which the risk is at an acceptable level, i.e., one that does not cause changes in the behavior of people and society [Kowalski 1997, p. 645, CIOP-BIP].

The literature distinguishes between the concepts of "work safety" and "occupational hygiene." The term "work safety" is related to the protection of life and means a set of conditions and technical, organizational, and economic measures in an organization that allow for the safe performance of tasks and the prevention of accidents at work [Ejdys et al. 2012, p. 136; Szłązak, Szłązak 2012, p. 11]. The term "occupational hygiene," on the other hand, is related to health protection [Ejdys et al. 2012, p. 136]. It is also a science that studies the impact of working conditions on the health of employees [Szłązak, Szłązak 2012, p. 11]. The term "occupational safety and hygiene" is understood as a set of conditions aimed at eliminating or completely limiting the negative impact of the work environment on the health of employees [Wróbel 2014, p. 93].

When analyzing definitions of occupational health and safety (OHS), it should be noted that most of them refer to the elimination of hazards in the workplace and the protection of the health and lives of employees. The most comprehensive definition identifies excessive physical and mental strain as a factor threatening employee safety. This definition is most relevant to occupational safety in the healthcare sector, where physical and mental strain on medical staff is a common phenomenon and particularly affects on-call staff [Rypicz 2020, p. 96].

HAZARDS IN THE WORKPLACE

Employers are obliged to provide their employees with safe and hygienic working conditions using appropriate scientific and technical achievements. There are a number of hazards in the work environment that affect occupational safety, including both tangible and intangible factors, such as physical factors, chemical substances and their mixtures, biological factors, and psychosocial/psychophysical factors. Physical, chemical, and biological factors are described as environmental factors, while psychophysical/psychosocial factors are described as organizational factors [Uzarczyk 2009, p. 13; Wieczorek 2014, p. 161; Wieczorek, Żukowski 2014, p. 63] (Figure 2.3).

All mental and physical work places a biological strain on the body and requires effort on the part of the employee. Therefore, in order to optimize it, it is necessary to create working conditions that counteract potential hazards in the work environment by ensuring the proper organization of workstations. This is the purpose of ergonomics, i.e., the science of finding the best ways and means to ensure optimal conditions in the

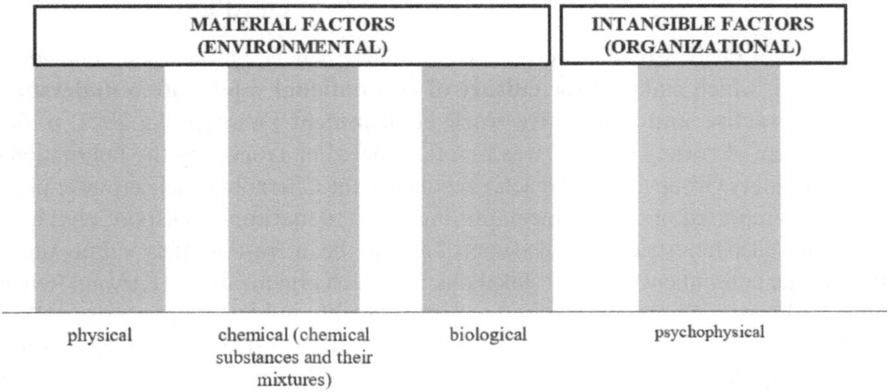

FIGURE 2.3 Classification of hazards affecting safety in the workplace

workplace [Witczak, Rypicz 2020, p. 77]. There are two trends in ergonomics: corrective and conceptual. *Corrective* ergonomics deals with the analysis of existing working conditions in terms of their adaptation to the psychophysical capabilities of employees and the formulation of recommendations, most often concerning the improvement of material factors in the environment. *Conception ergonomics*, on the other hand, deals with the design of workstations so that they ensure maximum work safety with the least possible physical and mental strain on the employee [Szlązak, Szlązak 2012, p. 234; Gałusza 2014, p. 426; Wieczorek, Żukowski 2014, p. 71; Rączkowski 2020, p. 253]. The task of ergonomics is to eliminate hazards at the stage of workstation design and to remove existing hazards, as well as to shape the right attitudes of employees toward the risk of accidents at work [Wieczorek 2014, p. 277].

One of the methods leading to improved OHS is **occupational risk assessment,** which provides feedback to the employer on compliance with legal requirements in the field of OHS. Occupational risk assessment allows the employee's exposure to hazards in the work environment to be estimated [Dudziak 2018, pp. 40–41]. It is used to identify circumstances in which occupational risk is at an unacceptable level requiring immediate intervention or at a medium level requiring measures to reduce that risk. The employer is obliged to develop an occupational risk assessment, which should be carried out periodically, as well as when introducing changes and creating new jobs, and after finding that protective measures are ineffective. This applies in particular to medical professions, where there is exposure to sharp medical instruments and biological agents [Wieczorek, Żukowski 2014, pp. 177–178; Dudziak 2018, p. 40].

The main objective of occupational risk assessment is to prevent occupational diseases and accidents at work [Gałusza 2014, p. 505]. Accidents at work are associated with loss of health and ability to work, suffering of the victims, and high economic costs for organizations and society. It is important to determine the causes and circumstances of accidents, which makes it possible to identify and eliminate hazards that were overlooked during occupational risk assessment, as well as to introduce preventive measures [Health and Safety Executive (HSE) 2004, p. 8; Rączkowski 2018, p. 329].

CREATING A CULTURE OF SAFETY IN THE WORKPLACE

The issue of OHS is related to the norms, values, beliefs, and learned behavior of employees, which make up **the culture of occupational safety, often understood as an attractive and supportive work environment** [Wiśniewska 2021, p. 68]. The concept of "safety culture" was first formulated in a report by the International Nuclear Safety Group after the reactor accident at the Chernobyl nuclear power plant. Accidents reported in the chemical, mining, rail and maritime transport, civil aviation, and space industries have also contributed to the interest in safety culture due to their tragic consequences for all stakeholders of such organizations [Kirwan, Reader et al. 2019, p. 200]. According to the British Health and Safety Executive (HSE), **"safety culture" is the result of individual and group attitudes, values, perceptions, competencies, behavior patterns, and the style and quality of safety management in an organization** [Stankiewicz, Sznajder 2010, pp. 16–17]. The literature presents several models of safety culture and one of them is the three-dimensional model of safety culture shown in Figure 2.4.

According to this model, safety culture is determined by three dimensions: human feelings, behaviors, and safety procedures within the organization. The strength of each dimension may vary, and each may influence the other two [Cooper 2000, p. 117]. In the context of healthcare, safety culture can be understood as an element of organizational culture that encompasses the entire organization, affects all professional groups, and consists of organizational and behavioral factors as well as the characteristics of the people performing the work.

Each organization has its own safety culture. Today, we distinguish between strong and weak safety cultures. A key factor in shaping a strong safety culture is the high value placed on health and life [Kim et al. 2016, p. 2]. The main goal of shaping **a strong safety culture** is to develop employee attitudes and behaviors that will protect their health and life as well as that of other employees. Creating a strong safety culture requires taking health and safety issues into account when planning and implementing organizational, technological, and personnel changes. Decisions made by organizations with a strong safety culture eliminate hazards and seek and implement solutions to improve workplace safety. All decision-making processes in these organizations prioritize safety over productivity and ensure the availability of personnel, procedures, and other resources necessary to ensure safety. Managers

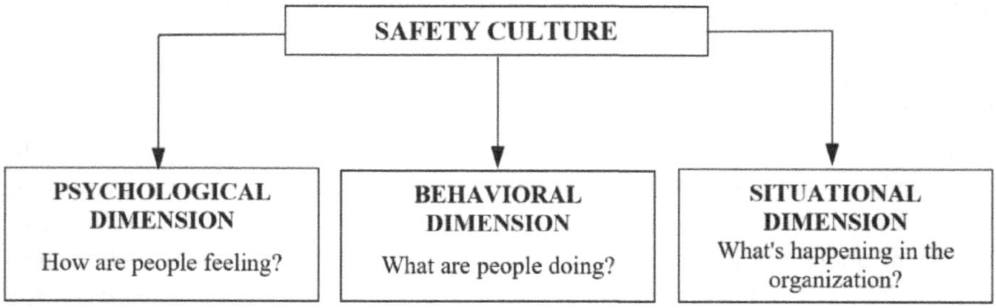

FIGURE 2.4 Dimensions of safety culture

have up-to-date knowledge of the aspects that determine workplace safety [Kim et al. 2016, p. 2]. Another factor in shaping a strong safety culture is the concern and commitment of senior management to health and safety issues. This factor is considered a fundamental aspect of effective safety performance [O'dea, Flin 2001, pp. 49–50; Rundmo, Hale 2003, p. 558; Nielsen 2014, p. 8]. In organizations with a strong safety culture, senior management demonstrates commitment, communicates information about accepted practices in the organization, sets rules of conduct and enforces their application, and promotes behavior patterns that promote safe work [Parker et al. 2006, p. 553]. A prerequisite for a strong safety culture is open communication based on mutual trust and a shared perception of hazards and risks at all levels of the organization [Nieva, Sorra 2003, p. ii18; Kirwan et al. 2019, p. 201]. Employees should feel comfortable communicating about safety issues, and supervisors should respond to information from employees and take action to eliminate or reduce the number of workplace accidents[1] and near misses.[2] An important feature of a strong safety culture is the non-acceptance of risky and dangerous behavior. In organizations with a strong safety culture, it is important to thoroughly analyze the causes of workplace accidents, learn from them, and implement changes to prevent such incidents in the future. A strong safety culture requires employee involvement and a sense of personal responsibility for safety issues on the part of each employee. Maintaining a strong safety culture also requires systematic analysis and assessment of occupational risks. A necessary condition for creating a strong safety culture is a focus on development and improvement that supports safety efforts. Another element of a strong safety culture is cooperation and collaboration between employees, as well as between employees and their supervisors. Good working relationships and relationships based on trust are important, especially when it comes to the use of preventive measures [Nieva, Sorra 2003, p. ii18; Kirwan et al. 2019, p. 201]. An important feature of a strong safety culture is the continuous monitoring of employee behavior and the promotion of safety-friendly attitudes. It is essential to motivate employees to initiate actions that improve health and safety and to reward such behavior and activities. The final prerequisite for developing a strong safety culture is job satisfaction, which is influenced by management style and leadership [Harvey et al. 2001, p. 618]. Table 2.1 summarizes the factors that illustrate the differences between a strong and a weak safety culture.

The differences between a strong and weak safety culture presented in the table show that the existence of appropriate documentation, procedures, and instructions does not guarantee appropriate health and safety practices. Conscious management of OHS is key here.

OHS management can be most simply defined as a way of managing occupational risk [Harms-Ringdahl 2004, p. 16]. It is an activity aimed at reducing the risk of loss of life and health at work to an acceptable level and maintaining it at that level or lower [Wieczorek 2014, p. 277]. It involves assessing the current state of OHS in an organization, developing and implementing measures to improve occupational safety by identifying hazards, implementing changes, evaluating the effects of implementation, and correcting activities. Issues related to OHS management are regulated by ISO 45001:2024. The standard is based on the analysis of risks and opportunities and takes into account the organizational context. It contains

TABLE 2.1
Determinants of Strong and Weak Safety Culture

Safety culture	
Strong	**Weak**
• High value placed on health and life • Open communication • Promotion of health and safety regulations and their consideration when implementing changes • No risky behavior • Elimination of hazards • Employee involvement in health and safety issues • Motivating employees to take action to improve health and safety • Systematic assessment of occupational risks • Monitoring employee behavior in terms of applicable health and safety rules • Training in occupational safety • Regular health and safety checks • Cooperation and collaboration between employees based on trust • Democratic leadership style • High level of job satisfaction	• Lack of interest in safety issues • Inappropriate actions by management and health and safety specialists • Encouragement of risky behavior • Failure to use collective and individual protective equipment • Low level of knowledge in the field of health and safety • Ignoring health and safety regulations • Equating risky work with high skills • Lack of mutual trust • Lack of standards and rules of conduct in the face of hazards • Lack of management responsibility for safe working conditions • Lack of intervention in cases of non-compliance with health and safety standards • Downplaying dangerous incidents

Source: Based on literature.

detailed requirements relating to workplaces: control, contractors, procurement, change management, outsourcing, process requirements, and input and output data for OHS management review [Pacana et al. 2017, p. 118]. The purpose of ISO 45001 is to support and promote good practices in OHS through the implementation of an occupational safety policy. The occupational safety policy includes information on responsibility for occupational safety and information on hazards and ways to keep them under control. It means a strategy for employees to deal with hazards, accidents, and occupational diseases [Szlązak, Szlązak 2012, p. 278].

The Safety Management System is used to manage OHS. It is a set of organizational measures aimed at ensuring conditions that allow work to be performed safely, without harm to health [Harms-Ringdahl 2004, p. 15; Szlązak, Szlązak 2012, p. 12]. The system allows for the organization of all activities in the area of OHS that result in improved working conditions. Effective OHS management requires: identifying hazards, determining the causes of errors made by employees, assessing risks and developing ways to reduce them, planning activities aimed at implementing the adopted safety policy, controlling working conditions, and motivating employees to behave safely in the workplace. An effective OHS management system reduces the number of accidents at work and occupational diseases, as well as the associated costs. It reduces employee absenteeism due to illness, increases productivity and improves the quality of work, helps to meet legal requirements, and has a positive impact on the economic performance of the organization [Pacana et al. 2017, p. 119].

SUMMARY

Technological developments, constant changes in the external environment, and emerging crisis situations have a strong impact on the specific nature of work in today's organizations. This necessitates constant monitoring of the work environment to identify emerging new hazards at work. Eliminating emerging risks in this area and ensuring the safety of employees in the healthcare sector helps to create a sustainable work environment, caring for the well-being of employees and patients, and thus achieving the organization's goals.

Creating a strong safety culture requires the implementation of an OHS management system and awareness of the risks associated with the high involvement of employees and senior managers in this area. Continuous training for employees, dialogue between managers and employees on health and safety issues, and the involvement of employees in practices that ensure safety at work are essential.

NOTES

1 An accident at work is considered to be a sudden event caused by an external factor resulting in injury or death that occurred in connection with work.
2 A near miss is a dangerous situation related to the work performed, during which no injury or deterioration of the employee's health occurs.

3 Hazards Occurring in the Healthcare Work Environment

The healthcare work environment is highly hazardous, and employees experience health and even life-threatening risks on an almost daily basis. This chapter classifies and discusses these risks. Particular attention is paid to psychosocial factors, whose negative impact is growing, including aggressive behavior by patients.

FACTORS AFFECTING WORK SAFETY

In every work environment, there are factors that can cause employees to experience reduced physical fitness, injuries, occupational diseases, or **other work-related illnesses**. Factors **present in the work environment** are classified according to their impact on the body and their properties.

Due to their impact on the body, factors are classified as dangerous, burdensome, and harmful. **Dangerous factors** are those whose effects can lead to immediate deterioration of health, accidents at work, or injuries. **Nuisance factors**, on the other hand, are those whose effects may reduce the ability to perform work and decrease productivity, but do not cause permanent deterioration of health. **Harmful factors** are those whose effects can lead to a gradual deterioration in health and then to disease [Wieczorek 2014, p. 169]. Depending on the level of impact and other conditions, an annoying factor can become harmful, and a harmful factor can become dangerous [CIOP-BIP[1]].

Depending on the nature of their action, **dangerous and harmful factors are divided into** [Uzarczyk 2009, p. 13; Wieczorek 2014, pp. 161–164; Dudziak 2018, pp. 45–62; Rączkowski 2020, pp. 173–174]:

- Physical, which include noise, mechanical vibrations, radiation, especially electromagnetic fields, lighting, temperature, electricity, aesthetics, air pollution, and sharp objects.
- Biological, which include microorganisms (bacteria, viruses), cell cultures, fungi, and parasites.
- Chemical, which occur in the form of vapors, gases, aerosols, liquids, or solids.
- Psychosocial/psychophysical, including organizational factors, management characteristics, and social and environmental conditions that may cause psychological, social, or physical harm to employees.

DOI: 10.1201/9781003716631-4

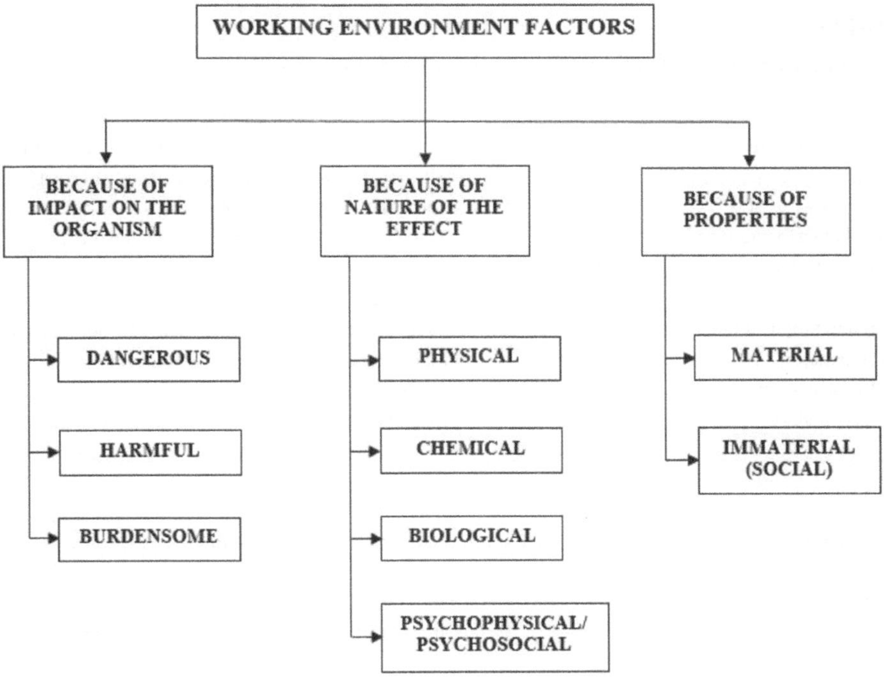

FIGURE 3.1 Classification of work environment factors

In turn, the classification of work environment factors according to their properties distinguishes between **material** and **immaterial (social) factors** [Jasińska 2010, pp. 255–256]. Material factors include physical, chemical, and biological factors. Material factors of the work environment are covered by health and safety regulations in the form of legal acts prepared and published by state authorities responsible for protecting the health and life of employees. Legal acts in individual countries contain established standards, compliance with which is the responsibility of the employer. An important group of factors occurring in the work process are non-material factors. They concern the functioning of the plant, work organization, and interpersonal relations [Szlązak, Szlązak 2012, pp. 81–83; Foldspang et al. 2014, p. 16; Wieczorek 2014, p. 167]. The classification of work environment factors is presented in Figure 3.1.

The classification of workplace factors presented in Figure 3.1 shows the types of hazards that employees may encounter while performing their work and which affect employees, negatively impacting their health.

PHYSICAL FACTORS IN THE WORK OF MEDICAL PERSONNEL

Physical factors affecting the safety of healthcare workers include: slippery surfaces, moving, carrying, transporting, inadequate lighting, noise, electromagnetic fields, radiation, gases, and sharp instruments. Physical factors [Rączkowski 2020, p. 173]

TABLE 3.1
Physical Factors Affecting Occupational Safety in Healthcare

Factor		Health effects
Manual handling of patients		Degenerative diseases and musculoskeletal disorders
Noise		Slowing down and discouraging work, and even provoking aggressive behavior
Mechanical vibrations		Vascular, nervous, and osteoarticular disorders
Electromagnetic field	Ionizing radiation	Risk of developing cancer
	Laser radiation	Eye diseases and skin damage
Lighting (optical radiation)		Fatigue, eye problems, and decreased work efficiency
Temperature		Malaise and decreased work efficiency
Electric current		Cardiac arrest and even death
Gas cylinders	Earth	Risk of explosion or fire and mechanical injuries
	Propane-butane	
	Medical	
	Aesthetic	
Sharp instruments		Puncture wounds and injuries

Source: Based on literature.

affecting occupational safety in the healthcare sector have been classified and organized in Table 3.1.

In the healthcare sector, there are hazards associated with the movement of people. These include steps, stairs, handrails, handles, balustrades, as well as wet, slippery, and uneven surfaces on which an employee may slip or fall. Such situations cause injuries among nursing and cleaning staff, laundry staff, as well as suppliers and ambulance personnel [European Commission (EC) 2011, p.116, p. 116]. Manual handling of patients, on the other hand, includes all activities during which a patient is lifted, pushed, pulled, moved, or carried. A European study of working conditions in the healthcare sector showed that 43.4% of workers have to lift or move patients and 27.7% have to move or shift heavy loads [European Commission (EC) 2011, p. 108]. According to the British Health and Safety Executive (HSE), more than 5,000 injuries to healthcare workers are reported worldwide each year, half of which are the result of manually moving patients [European Commission (EC) 2011, p. 108]. The study also found that 26.3% of workers suffer from back pain and 24.3% from muscle pain [European Commission (EC) 2011, p. 108]. Another hazard is noise, which is very annoying and whose effects are not immediately noticeable. Medical equipment emits various sounds, and it is difficult to clearly determine their impact due to the exposure of employees to other physical factors [Stansfield, Matheson 2003, p. 248]. A harmful physical factor to which healthcare workers are exposed is mechanical vibration generated by medical equipment and devices. Employees are most often exposed to local vibrations, which are transmitted through the upper limbs. The group of physical factors includes laser radiation, which is used for diagnosis and therapy as both an analgesic and anti-inflammatory agent. Laser treatments are also accompanied by elevated temperatures, which negatively affect well-being and work efficiency. In addition, laser treatments are accompanied by the emission of toxic compounds in the form of fumes, dust, and bioaerosols. One type of electromagnetic radiation is optical radiation: natural or artificial. When natural lighting is insufficient, employees use artificial light sources, especially in operating rooms and during night-time patient care. A significant hazard is electric shock, which can be caused by malfunctioning electronic equipment and faulty electrical installations supplying power to medical equipment and apparatus. Another hazard is propane-butane gas cylinders. Despite the growing trend toward automation, laboratories still use them in microbiological laboratories to sterilize small metal objects. Increasingly, hospital wards are equipped with analyzers for determining acid-base balance parameters, which are equipped with external calibration gas cylinders.

Sharp instruments such as needles, syringes, scalpels, glass ampoules (once opened), etc. are an integral part of the work of healthcare workers, especially in hospitals. Due to the constant flow of patients, especially in hospital emergency departments, and the performance of a range of medical interventions such as injections, intravenous infusions, blood sampling for laboratory tests, and minor surgical procedures, sharp medical instruments pose a particular risk to employees, as they can cause puncture wounds and injuries, and thus lead to diseases resulting from infection by biological agents[2] [PTEILCHZ[3] 2013].

BIOLOGICAL AGENTS IN THE WORK OF MEDICAL PERSONNEL

Harmful biological agents include "cellular microorganisms, internal parasites, non-cellular entities capable of replication or transferring genetic material, including genetically modified organisms, cell cultures that may cause infection, allergy, or poisoning." In the context of healthcare workers, three modes of infection are relevant:

- Through blood – microorganisms enter the employee's bloodstream as a result of contact with infected body fluids, usually blood or blood products – they pass through the mucous membrane or damaged skin.
- Through the air (droplets) – microorganisms are transmitted exclusively from one person to another, from human to human.
- Through direct and indirect contact, where direct contact involves the transfer of the pathogen from the skin or body surface of an infected person to another person – a patient or medical staff. Indirect contact, on the other hand, occurs through touching contaminated surfaces on which microorganisms are present.

The classification and list of harmful biological agents affecting the occupational safety of healthcare workers, including the route of infection, is presented in Table 3.2.

The European Union directive on the protection of workers from biological hazards classifies 379 agents into four groups (bacteria, viruses, fungi, and parasites) depending on the level of infection risk [Directive 2000/54/EC]. The table shows the biological agents that pose the greatest risk of infection to healthcare workers.

TABLE 3.2
Biological Agents Affecting the Occupational Safety of Healthcare Workers

Route of infection	Agent	Health effects
Blood	*Hepatitis B virus* (HBV)	HBV – can lead to cirrhosis of the liver and even death
	Hepatitis C virus (HCV)	HCV – asymptomatic for a long time. Later, weakness, muscle and joint pain, low-grade fever, and liver enlargement appear
	Human immunodeficiency virus (HIV)	AIDS – acquired immunodeficiency syndrome. The infection causes immune system disorders due to a decrease in the number of lymphocytes to levels that pose a risk of death
Airborne airborne droplet	Influenza viruses types A, B, and C (*Influenza virus*)	Fever, cough, runny nose, headache and muscle pain, weakness, and general fatigue
	Tuberculosis bacillus (*Mycobacterium tuberculosis*)	Causes lower respiratory tract disease, with symptoms including weakness, fever, cough, hemoptysis, shortness of breath, and chest pain
	Mononucleosis virus (EBV – *Epstein Barr Virus*) *Measles virus (MV)* *Rubella* virus (RUBV) *Varicella-zoster virus* (VZV)	Causes infectious diseases with varying symptoms and progression, ranging from malaise, fever, and skin changes to changes in internal organs
	SARS-CoV-2 (*Severe Acute Respiratory Syndrome Coronavirus-2*)	Causes COVID-19 disease, the symptoms of which include acute respiratory failure
Contact (direct and indirect)	*Methicillin-resistant Staphylococcus aureus* (MRSA)	Causes wound infections, boils, skin and organ abscesses, and sepsis
	Hepatitis *A virus* (HAV)	Causes hepatitis A (jaundice, damage to liver cells)
	Bacteria *(Shigella, Salmonella, Yersinia, Campylobacter, Mycoplasma, Chlamydia)*	Causes food poisoning, arthritis, diseases of the genitourinary system, skin and mucous membranes, and the eye

Source: Based on literature.

Of the biological agents listed, the most commonly observed infections are HBV and HCV infections, as well as HIV infections [Beltrami et al. 2003, pp. 168–175; EU-OSHA 2014, pp. 51–52]. These agents are not airborne, which means they pose a limited risk of infection to workers. They are dangerous, can cause serious illness in humans, and are highly likely to spread. Among the biological agents that pose a threat to the health of healthcare workers are MRSA and bacteria that cause respiratory infections, such as *Mycobacterium tuberculosis*. The risk posed by these factors is particularly high in places where elderly people are treated and where there is a high turnover of antibiotics, as well as in places where patients with tuberculosis are treated [KE 2011, p. 69].

Since 2020, the list of harmful biological agents has expanded. Currently, the work of medical personnel is burdened with an additional biological risk factor in the form of the SARS-CoV-2 virus. The SARS-CoV-2 virus has been classified as a biological agent that can cause serious illness in humans, is dangerous to workers, and is highly likely to spread among the human population. The virus is transmitted through droplets, causing a disease called COVID-19, which can lead to pneumonia and acute respiratory distress syndrome. In January 2020, the World Health Organization declared a public health emergency of international concern and then classified the disease as a pandemic. The circumstances of the epidemic significantly changed working conditions and perceptions of safety. Public health recommendations changed from day to day. This resulted in an increasing workload for staff, not only due to the growing number of patients suspected of being infected with SARS-CoV-2 and the need to care for those recovering, but also due to staff infection and quarantine. In addition, caring for patients while wearing personal protective equipment (coveralls, shoe covers, goggles, face shields, masks, two pairs of gloves) limited manual dexterity and made work significantly more difficult. This situation led to an increase in the risk to life and health and a deterioration in the working environment for medical staff.

CHEMICAL FACTORS IN THE WORK OF MEDICAL PERSONNEL

Chemical factors are another factor present in the working environment of medical personnel. Chemical factors present in the work process have been divided according to their effect on the human body and according to their methods of absorption. Chemical factors present in the work process of healthcare workers and the effects caused by their action are listed in Table 3.3.

Chemical agents that pose a risk to healthcare workers include substances and mixtures used in disinfection and sterilization processes, as well as detergents. A European study of working conditions has shown that people working in the healthcare sector (doctors, nursing staff) are more exposed to biological and chemical hazards because they often come into contact with infectious materials and chemicals used to disinfect instruments and workplaces [de Jong et al. 2014, p. 8]. Another hazard for healthcare workers is latex. Contact with latex can trigger a number of symptoms, including contact urticaria, rhinitis, conjunctivitis, shortness of breath, and complications in the form of anaphylactic shock. A chemical that poses a hazard to healthcare workers is water. Although it is colorless, odorless, and tasteless, it contains dissolved and insoluble substances: hydrocarbons, resins, fats, proteins, glass, and metals, and even gold and silver. In addition to detergents and metals, it may contain radioactive substances, dyes, fertilizers, and pesticides. Cleaning staff who clean sanitary facilities and disinfect surfaces and beds are exposed to the increased effects of water [KE 2011, p. 231]. Both hand washing and disinfection play a key role in infection prevention. They protect against infection with pathogenic microorganisms, limit their transmission to patients, and minimize the risk of spreading infections [Rosiński et al., 2019, p. 2]. Careful washing and disinfection of the skin

TABLE 3.3
Chemical Factors Affecting Occupational Safety

Agent			Health effects
Detergents, disinfectants, and sterilants			Flammable agents with corrosive or sensitizing effects. They affect the skin, mucous membranes, and respiratory tract, and may impair reproductive functions.
Latex			Allergic reaction, causes skin irritation and damage
Water			Potential irritant. Causes swelling and shrinkage of the outer layer of the epidermis, changes in the form of redness, swelling, cracks and peeling, as well as itching and dermatitis, known as contact dermatitis
Chemical compounds (medicines, cytostatics)			Disorders of well-being: headaches and dizziness, as well as local reactions: allergies, irritation (redness, burning, itching), tissue destruction (necrosis). They have toxic, mutagenic, carcinogenic, or teratogenic properties.
Gases	Medical	Oxygen	Fire and cylinder explosion due to increased ambient temperature (heating)
		Carbon dioxide	Circulatory failure, headaches, nausea, and vomiting, which may lead to loss of consciousness and even death
	Aesthetic	Nitrous oxide Halothane Enflurane Isoflurane Sevoflurane Desflurane	Harmful effects on a single organ or bodily function, or toxic or allergic effects. They cause fatigue, headaches, mood swings, reduced neuropsychological performance, reduced fertility, and complications during pregnancy.

Source: Based on literature.

surface of the hands can remove approximately 90% of bacteria [Dzierżanowska-Fangrat et al. 2010, p. 5]. Hospital staff come into contact with chemical compounds that are components of medicines. These can be absorbed through the respiratory tract and through intact skin, thus entering the bloodstream and impairing the functioning of the body. Important chemical factors commonly used in medicine are medical and anesthetic gases, which have an adverse effect on the health of employees [European Commission (EC) 2011, p. 241].

In addition to the physical, biological, and chemical hazards described above, psychosocial/psychophysical factors should also be noted, which are listed in the next subsection.

PSYCHOSOCIAL FACTORS IN THE WORK OF MEDICAL PERSONNEL

Psychosocial factors are also referred to as psychophysical factors. The difference is that psychosocial factors cause stress on the mental sphere of a person, while psychophysical factors cause stress on the physical sphere [Rączkowski 2020, p. 425]. In practice, the term psychosocial factors is used to indicate that their effect is a mental and physical strain on the body (work overload) [Dudziak 2018, p. 62]. The International Labor Organization (ILO) defines **psychosocial factors as** the interaction between the content of work, the management and organization of the work process, and the capabilities of workers, which can affect workers' health, productivity, and job satisfaction [Leka, Jain 2013, p. 9; ILO (International Labour Office) 2017, p. 7]. This approach points to the aforementioned imbalance between employees' resources (e.g., their abilities, skills) and the challenges they face on a daily basis in the workplace. It is therefore closely linked to employee well-being in the workplace. It is no surprise that psychosocial factors are also defined as those aspects of work design and management, as well as the social and organizational contexts of work, that **can cause mental or physical harm** [ILO (International Labour Office) 2017, pp. 8, 13] or those that **affect the well-being and health of employees** [Leka, Jain 2013, p. 16] and occupational safety and health [ILO (International Labour Office) 2017, p. 7]. Psychosocial factors can therefore be understood as working conditions that affect an employee's life situation, physical and mental health, and personal and professional satisfaction. The changing work environment gives rise to new risks, and the definition of psychosocial risks may therefore evolve. A summary of psychosocial/psychophysical factors affecting the occupational safety of healthcare workers and their characteristics is provided in Table 3.4.

It is impossible to identify all the psychosocial factors present in the working environment of medical personnel. Only those that have the greatest impact on the lives of healthcare workers are presented here. Due to the negative consequences of psychosocial factors, it seems reasonable to discuss them in more detail.

Among psychosocial factors, static physical strain is particularly severe, resulting from prolonged muscle tension caused by holding the body or objects in a forced position for long periods of time. This is facilitated by the ergonomics of buildings, insufficient space to perform work, inadequate workplace layout, and excessive working hours without breaks. A forced body position (standing), often combined with bending, characterizes the work of doctors and nursing staff. Some medical staff work in a forced body position – sitting while working with medical records in combination with working at a computer monitor. Another risk factor in the psychosocial group is work overload and time pressure. These result from a shortage of medical staff and underfunding of health services in some countries.

Staff shortages can result in long hours of work without breaks for other medical staff, leading to disturbances in the proper functioning of the body, high levels of physical and mental fatigue, and a high risk of errors or accidents at work. Studies have shown that long working hours impair bodily functioning to the same extent as a blood alcohol concentration of 0.05 mmol/L [Dawson, Reid 1997, pp. 235–237]. *The International Organization for Migration* has pointed to evidence that nursing staff working more than 12 hours in a 24-hour period or more than 60 hours in a 7-day period, and not having sufficient breaks, slows reaction times, reduces concentration, and is the most common cause of errors [Trossman 2007]. The risk of medical error is twice as high after 12 hours of work than during 8 hours of work [Rypicz 2020, p. 93]. Medical staff belong to a group of employees who perform their professional duties in a variable circadian rhythm. They work nonstandard hours, often starting work very early in the morning until late in the evening

TABLE 3.4
Psychosocial Factors Affecting Occupational Safety

Factor	Characteristics
Job content	The work is characterized by a high level of uncertainty, the need for increased concentration, constant exposure to other people during work, uncertainty about the effects of work, and a lack of a sense of security.
Physical strain (static)	It is most severe when forced to stand without the possibility of changing to a sitting position (performing long surgical operations). The work requires prolonged muscle tension.
Work overload, time pressure	The work is team-based, emotionally demanding, requires considerable mental effort and readiness to react quickly, as well as appropriate qualifications. It requires constant vigilance, even when working at night. In addition, it is characterized by a forced pace, considerable time pressure and pressure from other people, and the rhythm of work is often imposed.
Form of employment (contract agreement)	Favors long-term work, even for several days in a row with short breaks.
Work schedule (work in a variable daily rhythm)	The work is shift-based, takes place at night, and is not adapted to social rhythms. It is characterized by fixed working hours, long daily working hours, and requires availability.
Shortage of medical personnel	This is due to too few positions and a large number of employees of retirement age. In addition, the shortage is caused by the widespread migration of medical personnel: doctors, nursing staff, midwives, and paramedics.
Patient expectations	Patients expect respect, individual treatment, interest, explanations of the causes and effects of their illness, emotional support, and information about tests and treatment, which is why communication and building good relationships are very important.
Control over work	The work is characterized by a lack of influence on physical working conditions, workload, pace of work, and work style. Employees have little say in decision-making processes and cannot interrupt their work when necessary.
Interpersonal relationships	These refer to relationships at home, with colleagues, patients, and their unpredictable reactions. Poor relationships lead to exposure to aggression from patients and their families, intimidation, harassment, shouting, insults, blackmail, and threats.
Personal considerations	Lack of skills, experience, physical and health predispositions, insufficient intellectual and mental predispositions

Source: Based on literature.

or starting work in the evening until the early hours of the morning. Night work is particularly stressful, leading to changes in circadian rhythms and the development of *shift lag (industrial jet lag)* [Zużewicz 2017, p. 130]. Night work causes chronic fatigue, reduced muscle strength, impaired concentration and hand-eye coordination, a change in work pace from fast to slower (slowed reaction time), reduced productivity, and a feeling of inability to cope with work. Studies have shown that employees' productivity is at its lowest in the morning, immediately after reaching their minimum body temperature [Rypicz 2020, p. 93]. In addition, night work results in sleep disorders and digestive and circulatory problems [Rypicz 2020, pp. 92–93]. A study of the health of nursing staff showed a 1.4-fold increase in the risk of non-fatal ischemic heart disease and a 1.2-fold -fold increase in the risk of fatal ischemic heart disease among women who had ever worked in a shift system involving night work, with an increased risk of ischemic heart disease already observed after 6 years of shift work [Leka, Jain 2013, p. 43]. Studies on the differences between day and night shift workers have shown that night shift workers are more likely to experience reduced control at work, low social support from supervisors, and physical and psychological violence [Leka, Jain 2013, p. 46].

Another risk factor in the psychosocial category is the degree of control over one's work. A lack of control over one's work is common in the healthcare sector, as confirmed by the ESENER survey[4] conducted by the European Agency for Safety and Health at Work (EU-OSHA). Work in healthcare is performed on the basis of legal regulations (laws, regulations, legal provisions), standards, guidelines, medical procedures (rules of conduct), and recommendations for the provision of medical services. The organization of work is also determined by regulations and guidelines from superiors, which employees are required to follow. A lack of or low level of control over work is associated with stress, anxiety, depression, apathy, exhaustion, low self-esteem, and an increased risk of cardiac symptoms [Leka, Jain 2013, p. 51].

Another threat classified as a psychosocial factor is **patient expectations**, which cannot always be met. Patients want immediate access to a given medical service, which is often impossible, resulting in patient dissatisfaction in the form of complaints and claims, as well as socially unacceptable behavior toward medical facility staff.

In healthcare facilities, the relationship between medical staff and the patient and their family is very important, as it determines the assessment of the quality of services provided. **The relationship between the patient and the medical staff** is of a special nature. On the one hand, these relationships are shaped by patient rights [relevant laws in individual countries], compliance with which is a statutory obligation of medical professionals and other persons in contact with patients. These relationships involve cooperation with patients and their families in formulating diagnoses and planning treatment. On the other hand, the relationship with the patient is informal, in which the staff strives to establish a good relationship with them. Establishing good relations with patients is difficult due to their lack of medical knowledge and difficulty in understanding the specialized information provided by medical staff, as well as the need to entrust their health and lives entirely to doctors, nursing staff, and paramedics. It is becoming increasingly common for interpersonal relationships between patients and medical staff to be characterized by negative associations. These are expressed through aggressive behavior on the part of patients, both physical and verbal, in the form of threats, shouting, insults, and intimidation [cf. Kuriata-Kościelniak 2020; Pallarés et al. 2023; Rabin et al. 2024, Tarczoń et al. 2025].

Exposure to psychosocial factors can lead to negative effects, among which occupational stress is at the forefront. Occupational stress in healthcare facilities occurs relatively often due to the large number of factors that determine its occurrence. NIOSH defines **occupational stress** as harmful physical and emotional reactions that occur when job demands exceed an employee's capabilities [NIOSH 2008[5]]. In general, occupational stress can be defined as the body's response to demands that exceed its ability to cope. It occurs when an employee's competencies are not matched to the demands of the job.

Healthcare facilities experience high levels of occupational stress caused by work overload, pace of work, shortage of medical staff, and patient expectations [Van den Tooren, De Jonge 2010, p. 40]. Another cause of stress is the enormous responsibility for the lives of others and the decisions made, the great diversity and changing situations that require quick decisions and different courses of action, as well as constant contact with patients and death [Gałusza 2014, p. 274]. These stressors vary depending on the profession and the tasks performed. Studies have shown that common stressors in the work of nursing staff include work overload, time pressure, rushing, lack of social support, especially from senior management, and exposure to infectious diseases. Other stressors include needle stick injuries, lack of sleep, staff shortages, career development issues, dealing with difficult or seriously ill patients, limited ability to help some patients, and exposure to violence or threats. This "catalog" of stressors includes low social evaluation of work and work schedule (shift work, unpredictable working hours, and inflexible work schedules) [Bartczak, Bartczak 2010, p. 11]. Research conducted during the SARS-CoV-2 epidemic has shown that higher levels of occupational stress among nursing staff are caused by working in direct contact with patients infected with SARS-CoV-2 [Piotrowski et al. 2021, p. 462]. The consequence of these factors is a reduction in the efficiency and quality of work performed. In turn, studies conducted among doctors have shown that stress factors include long working hours (medical shifts), excessive workload, and lack of control over the pace of work, which is a result of a shortage of medical staff and too many patients per medical staff member. The responsibility for the lives and health of patients has a negative impact on the health of doctors. Dealing with death, interpersonal conflicts with other employees, patient expectations, and the threat of lawsuits for malpractice are particularly difficult [NIOSH 2008[6]]. Furthermore, there is a great deal of diversity and a changing situation that requires quick decision-making and different courses of action [Gałusza 2014, p. 274]. In turn, the results of research on the assessment of stress at work in a group of paramedics showed that their work is stressful, dangerous, physically and mentally exhausting, and requires special personality traits. It is interesting, appreciated by loved ones and colleagues, but underappreciated by superiors, and additionally has a negative impact on family and private life. The quality of patient care can be a stressful risk factor for people in the medical professions [NIOSH 2008[7]]. The literature on the subject also indicates that personal factors, including a hostile attitude toward work, insufficient experience, and limitations and contraindications to performing a specific job, may be a significant factor threatening safety at work [Uzarczyk 2009, p. 16].

SUMMARY

The healthcare work environment is considered one of the most dangerous. Employees are exposed to all possible hazards, including physical, chemical, biological, and psychosocial ones. Despite technological progress and civilizational development, the sources of risk at work in this sector have increased in recent years. During the SARS-CoV-2 pandemic, many factors threatening the safety of healthcare workers have been exacerbated, especially psychosocial factors such as high demands, overtime, excessive work pace, insufficient time for recovery, unsatisfactory pay, and occupational stress [Ilczak et al. 2021, pp. 139–149; Somville et al. 2021, pp. 373–383]. The European Agency for Safety and Health at Work adds to these factors working with demanding clients – patients in the case of healthcare [EU-OSHA 2023], which is becoming increasingly stressful. This is particularly true in the context of violence and aggressive behavior by patients, which determines the safety of healthcare workers. An analysis of the literature has revealed a lack of research in this area, particularly with regard to the links between manifestations of violence and aggressive behavior by patients and medical staff: doctors, nursing staff, and paramedics. Authors of books on occupational health and safety describe the hazards present in the work environment, but do not address the issue of aggressive behavior by patients toward medical staff or devote little attention to it [Witczak, Rypicz 2020]. Available studies [Gillespie et al. 2010; Aljohani 2021; Merecz, Mościcka 2003; Suwała 2008] refer to the phenomenon of violence, which is equated with aggression and used interchangeably, making it difficult to interpret the phenomena correctly and assess their scale.

NOTES

1. CIOP-BIP, *Dangerous, harmful, and burdensome factors and their consequences in micro and small retail businesses* (https://m.ciop.pl/CIOPPortalWAR/appmanager/ciop/mobi?_nfpb=true&_pageLabel=P41800156714967427910298&html_tresc_root_id=300007035&html_tresc_id=300007030&html_klucz=300006222&html_klucz_spis 27.04.2021).
2. Issues related to biological agents in the workplace are regulated by Directive 2000/54/EC of the European Parliament and of the Council of September 18, 2000, on the protection of workers from risks related to exposure to biological agents at work.
3. The abbreviation PTLECHZ stands for: Polish Society of Epidemiologists and Infectious Disease Physicians.
4. ESENER is a European survey of enterprises on new and emerging risks. The survey is conducted periodically and its aim is to identify incentives and barriers affecting occupational safety management in enterprises, with particular emphasis on psychosocial factors related to work.
5. NIOSH (National Institute for Occupational Safety and Health), Department of Health and Human Services Centers for Disease Control and Prevention, 2008, *Exposure to Stress,* Occupational Hazards in Hospitals (https://www.cdc.gov/niosh/docs/2008-136/pdfs/2008-136.pdf 07.05.2021).
6. NIOSH (National Institute for Occupational Safety and Health), Department of Health and Human Services Centers for Disease Control and Prevention, 2008, *Exposure to Stress,* Occupational Hazards in Hospitals (https://www.cdc.gov/niosh/docs/2008-136/pdfs/2008-136.pdf 07.05.2021).
7. NIOSH (National Institute for Occupational Safety and Health), Department of Health and Human Services Centers for Disease Control and Prevention, 2008, *Exposure to Stress,* Occupational Hazards in Hospitals (https://www.cdc.gov/niosh/docs/2008-136/pdfs/2008-136.pdf 07.05.2021).

4 The Different Faces of Violence and Aggression in Healthcare

Violence and aggression in the workplace pose a serious threat to employees, causing many negative consequences for both employees and the organization itself. In the healthcare sector, employees are increasingly experiencing aggression from patients. This chapter presents an analysis of existing publications on the issue of occupational safety in the healthcare sector.

The literature was analyzed using a systematic review of the subject, supplemented by a review of websites. The search was conducted using a mixed strategy. The electronic search was carried out using online databases: Scopus (Elsevier), Medline/PubMed, Springer, using keywords such as: *Healthcare, Emergency Department, working environment, physical, chemical, biological and psychosocial threats, violence and aggression in Healthcare*. The databases were selected due to their extensive collection of current publications in the fields of management, as well as medicine. Full-text, peer-reviewed publications available in scientific journals were included. The analysis was then narrowed down to publications related to the study of occupational safety by adding the term "occupational safety" searched for in the title of the publication. The websites of key occupational safety organizations were also searched, such as: the European Agency for Safety and Health at Work, the British Health and Safety Executive, and the US National Institute for Occupational Safety and Health. The following keywords were used: *health and safety, risk factors, violence and aggression in the workplace.*

AGGRESSIVE BEHAVIOR OF PATIENTS

Explaining aggressive behavior prompts an analysis of the most important theories explaining the mechanisms of such behavior [Krahé 2005, p. 31]. Table 4.1 presents theories that show how different the mechanisms triggering aggressive behavior are and how many of them can be supported by examples of patient behavior toward medical staff. The presented theories of aggression will serve as an introduction to examples of aggressive behavior by patients in healthcare facilities.

One of the many theories that explains the mechanism triggering aggression is the body type theory formulated by William Herbert Sheldon. The theory is based on measurements of body proportions. The author identified three types of human body types: ectomorphic, endomorphic, and mesomorphic [Kumięga 2014, p. 44]. The easiest to recognize is a person with an ectomorphic body type. This is usually a person with a slender figure with relatively low muscle mass, but with a well-developed

DOI: 10.1201/9781003716631-5

TABLE 4.1
Summary of Selected Theories of Aggression

Author	Theory	Characteristics
Sigmund Freud (1927)	Psychological (*catharsis*)	The theory suggests that human behavior is governed by two forces that are in constant conflict with each other. One is the instinct for life and love, and the other is the death instinct, which causes tension and stimulates its release.
John Dollard (1939)	Frustration	According to the theory, there is a link between frustration and the emergence of aggressive behavior. Frustration caused by the inability to realize one's life plans triggers aggressive behavior.
William H. Sheldon (1940)	Body type	According to theory, body type influences the tendency toward aggressive behavior. The author distinguished three body types, each of which influences aggressive behavior in a different way.
Neal E. Miller (1950)	The influence of rewards and punishments	According to the theory, aggression results from the influence of rewards and punishments, and the factor that inhibits aggressive behavior is the awareness of the expected punishment.
Arnold H. Buss (1961)	Habit	According to this theory, aggression is a habit and can be a characteristic feature of a person's personality. The concept is similar to the theory of rewards and punishments.
Leonard Berkowitz (1962)	External and internal factors	The theory assumes that aggressive behavior depends on a person's internal readiness and external stimuli. The interaction of these two factors is a necessary condition for the emergence of aggressive behavior.
Konrad Lorenz (1963)	Instinct (instinctive behavior)	According to the theory, aggression is one of four instincts, alongside hunger, flight, and sexual drive. The author believed that instinct is innate and its energy must be discharged, causing aggressive behavior.
Albert Bandura, Richard Walters (1965)	Social learning	The theory suggests that aggressive behavior is the result of imitation and personal experience. People learn aggressive behavior by observing others who behave aggressively.

Source: Based on literature.

central nervous system. The opposite of an ectomorph is a person with an endomorphic body type. An endomorph is a person of short or medium height, prone to accumulating fat, stocky, corpulent with poorly visible muscles, characterized by mood swings. A mesomorph, on the other hand, is a combination of the characteristics of an ectomorph and an endomorph. People with a mesomorphic body type are usually strong and muscular, active and aggressive in their actions. According to Sheldon, people with a mesomorphic body type are most prone to aggressive behavior [Kumięga 2014, p. 44]. An example that may explain Sheldon's theory is the behavior of a patient with a short stature and stocky build, with abdominal obesity, accompanied by mood swings.

Another approach to aggression was proposed by Sigmund Freud, who believed that human behavior is driven by two opposing instincts, one of which leads to destructive actions. According to the author, humans possess energy that must be used and discharged regardless of the situation. According to Freud, when a person is unable to directly attack the source of aggression, they attack an object that they believe is less likely to be punished. This situation is referred to as displacement of aggression [Kumięga 2014, pp. 44–45]. An example of patient behavior that explains Freud's concept is anger toward a doctor, discharged on another member of the medical staff. Another example is the change from direct to indirect aggression, in which the patient slams the door instead of using physical aggression toward the medical staff.

Another concept explaining the emergence of aggression is the theory of instinct (instinctive behavior) formulated by Konrad Lorenz. According to the author, no "special" conditions or "exceptional" situations are needed to release energy [Kumięga 2014, p. 45]. An example of patient behavior that explains the assumptions of instinct theory is a spontaneous aggressive reaction without a specific reason.

Another concept that may explain aggressive patient behavior is the frustration theory proposed by John Dollard. According to the author, aggression is the result of a drive to release frustration, where frustration is understood as a disruption of goal-oriented behavior. According to Dollard, the force of frustration increases in direct proportion to the number of failures that prevent the intended goal from being achieved. The greater the number of failures, the stronger the frustration and the greater is the likelihood of aggressive behavior [Kumięga 2014, p. 45]. An example would be a patient who, due to medical conditions, must be hospitalized and cannot pursue their life plans. Frustration builds up in them, which often causes aggressive behavior directed at medical staff.

Neal E. Miller reformulated John Dollard's concept, arguing that frustration triggers a range of behaviors in humans, and aggression is only one of them. The author argued that aggression, like motivation, results from the influence of rewards and punishments, and that it disappears over time and is associated with a change in the object of aggression or the form in which it is expressed [Kumięga 2014, p. 50]. An example illustrating the assumptions of this theory in a healthcare facility would be a situation in which a patient fears that they will not receive medical assistance from a doctor (treated as a punishment). To avoid punishment, they change the object toward which they express their aggressive behavior (the victim) or change the form of expressing aggression by using jokes and gossip.

The etiology of aggressive behavior is also explained by Leonard Berkowitz's theory of external and internal factors. The author assumes a close relationship between external and internal factors. The stronger the internal stimuli an individual has, the weaker the external stimuli needed to manifest aggressive behavior. Conversely, the weaker the internal stimuli, the greater the need for external stimulation, e.g., more violent provocations [Kumięga 2014, p. 46]. In healthcare, an example that explains Berkowitz's theory is aggressive patient behavior caused by internal factors such as mental illness, excessive consumption of alcohol and/or psychoactive substances, or addiction to them, and by external factors such as a difficult life situation or the presence of accompanying persons (family members, partners).

Aggressive patient behavior can also be analyzed from the perspective of Albert Bandura and Richard Walters' *Social Learning Theory*. According to social learning theory, aggressive behavior is influenced by the upbringing process. Depending on whether aggressive behavior was supported or condemned, it may appear or disappear later in life. The theory also draws attention to the influence of the media, which, by showing aggressive behavior, causes desensitization and a greater tendency to exhibit such behavior. Observing and modeling aggressive television characters or computer game characters reduces sensitivity and becomes a source of ideas for releasing aggression, making it easier and quicker to display [Kumięga 2014, p. 52]. In the healthcare sector, an example confirming the assumptions of this theory is the behavior of a patient who witnesses the aggressive behavior of another patient who has resolved a conflict by using force or has achieved a favorable outcome as a result of aggressive behavior.

Aggressive behavior in patients can also be explained by the habit theory formulated by Arnold H. Buss. The author suggests that people exposed to factors causing anger, rage, nervousness, and frustration behave more aggressively than people who are not exposed to such factors. According to Buss, human behavior is modified by external factors such as rewards and punishments. The author believes that rewarded behaviors are likely to be repeated, while punished behaviors will be significantly weakened. Similarly, aggressive behaviors that are not reinforced by rewards will be significantly weakened, while those that are rewarded will be reinforced (according to Skinner's concept). An example would be the behavior of a patient who has adopted aggressive behavior from their parents, siblings, or peers, or who has behaved aggressively for some time and now does so automatically.

Individual factors such as personality traits, attitudes, values, and illnesses such as schizophrenia, depression, endocrine disorders, and post-traumatic stress disorder also play a significant role in the etiology of aggressive behavior [Kumięga 2014, p. 48] and insufficient knowledge of ethics [Mikos et al. 2018, p. 58]. In addition, aggressive behavior can be triggered and intensified in people who are delirious, intoxicated with alcohol or psychoactive substances, or experiencing withdrawal symptoms [Lickiewicz 2017, p. 15; Mikos et al. 2018, p. 58]. Alcohol is not directly associated with aggressive behavior, but it reduces self-control and resistance to actions that violate basic social and cultural norms. It reveals desires and aspirations that are limited by social and moral norms [Kumięga 2014, p. 46].

VIOLENCE AND AGGRESSION IN THE WORKPLACE

An analysis of the literature on violence and aggression reveals a number of problems concerning the definition and description of these phenomena. Due to the ambiguity of the term "**violence**," the definitions available in the literature are summarized in Table 4.2.

The definitions in the table have common elements such as the use of physical force, causing harm or damage, and affecting health. Some authors define "violence" more narrowly as an attack on another person [Perrone 1999]. Others understand the concept broadly, indicating that it is influencing a person's thought processes,

TABLE 4.2
Summary of Definitions of Violence

Author	Definition
S. Perrone (1999)	An act of aggression intended to cause extreme harm
J. R. Ackerman, S. Pickering (2002)	Behavior that humiliates an individual, causing physical injury and thereby violating the law
World Health Organization (*World Health Organization – WHO*) (2002)	The deliberate use of physical force against oneself, another person, a group, or an entire community, which causes or is highly likely to cause injury, death, prejudice, maldevelopment, or deprivation
A. Pilszyk (2007)	The use of force or power to harm those who are weaker
European Commission (EC) (2011)	Incidents involving the abuse, intimidation, or assault of individuals in the workplace, posing a threat to their safety, health, and well-being
D. Grudzień, P. Zurzycka, T. Radzik (2015)	The deliberate use of physical force against oneself, third parties, or a group of people. An act of aggression intended to cause physical pain or humiliation
A. Stępień (2015)	Human behavior that causes physical harm to another person or acts intended to cause such harm
J. Helios, W. Jedlecka (2017)	A relationship between people based on the use of overwhelming force against another person. Its purpose is to cause harm to the victim, inflict pain, cause suffering, or humiliate them
J. Lickiewicz (2017)	An attack on another person with or without the use of dangerous objects. More broadly, influencing a person's thought process, behavior, or physical condition without their consent

Source: Based on literature.

behavior, or physical condition without their consent [Lickiewicz 2017]. For the analysis of violence in the workplace, the definition of the European Commission (EC) was adopted, which best describes unacceptable behavior in the workplace that poses a threat to the safety and health of employees. Insults, threats, and physical or psychological aggression directed at a person performing work, including by persons outside the organization, are considered violence [European Commission (EC) 2011, p. 187].

When analyzing the relationship between violence and aggression, it is worth examining the definitions of aggression. **Aggression** is most often understood as deliberate behavior aimed at harming another person, injuring them, or causing them pain [cf. Perrone 1999; Anderson, Bushman 2002; Kenrick, Neuberg, Cialdini 2002]. Some authors add that these hostile behaviors cause loss or harm to another person [Grudzień, Zurzycka, Radzik 2015; Stępień 2015]. In turn, Lickiewicz, in defining aggression, points out that the person who is the subject of aggression is motivated to avoid it [Lickiewicz 2017].

The elements that recur in the definitions of aggression presented in the literature are: deliberate behavior, causing harm or injury, inflicting suffering or pain. A comparison of both definitions shows that authors use the terms violence and aggression interchangeably. The phenomena of violence and aggression are classified according to various criteria, hence a number of types of these phenomena can be found in the literature on the subject, which may overlap.

Most often, physical and psychological violence are distinguished [Helios, Jedlecka 2017, pp. 18–20]. **Physical violence** includes violations of physical integrity that cause pain or bodily harm, and its forms include kicking, choking, burning, pushing, slapping, and even the use of weapons [Helios, Jedlecka 2017, pp. 18–20]. Physical violence can result in fractures, burns, contusions (bruises), and cuts [Kmiecik-Baran 1998, p. 370].

Psychological violence, also known as **emotional violence**, includes situations in which the perpetrator exercises psychological control over the victim. Forms of psychological violence include humiliation, ridicule, manipulation, criticism, prohibition, threats, name-calling, and embarrassment [Kmiecik-Baran 1998, p. 370]. Psychological violence also includes intimidation, threats, blackmail, and destruction of private property [Brown, Herbert 1999, pp. 29–30]. Depending on the intentions, a distinction is made between spontaneous (hot) and cold violence [Helios, Jedlecka 2017, p. 29]. The source of spontaneous violence is frustration. This type of violence is characterized by anger, rage, and fury; it appears quickly and disappears quickly. It is accompanied by shouting, name-calling, insults, physical violence, impulsive behavior, and the violent infliction of pain [Helios, Jedlecka 2017, p. 29]. Cold violence, on the other hand, occurs where there is an element of cold calculation. Its purpose is to achieve certain positive values or goals [Helios, Jedlecka 2017, p. 29]. Another classification distinguishes instrumental violence, which occurs directly in interpersonal relationships and is recognizable only by those affected [Helios, Jedlecka 2017, pp. 29–30]. This type of violence can occur in the healthcare sector, especially in the sphere of contact between medical staff and patients.

The literature offers yet another classification of violence: violence directed against oneself (self-aggression), interpersonal violence, and collective (group) violence. Violence directed against oneself includes self-harm and suicidal behavior, while collective violence refers to violence of a political, social, and economic nature. Interpersonal violence, on the other hand, most often occurs in the family and where the victims may be children, partners, the elderly, and relatives. This type of violence can also occur in healthcare facilities due to the fact that interpersonal relationships are an important factor in the work environment [Krug et al. 2002, p. 6; Tobiasz-Adamczyk 2014, pp. 7–8]. The literature also contains other classifications of violence that are not reflected in incidents occurring in the healthcare sector and therefore have not been cited here.

Healthcare workers are particularly exposed to physical and psychological violence [Suwała 2008, p. 158]. Over 25% of acts of violence reported in the workplace occur in the healthcare sector, where there is a high level of stress, workload, high demands and pace of work, contact with patients, and interactions with their families [Suwała 2008, p. 158]. Violence in healthcare facilities can also have a negative impact on the functioning of the organization, reducing the quality of services provided and causing employee turnover, which, in the context of staff shortages, can further limit patients' access to medical services.

In order to present a complete picture of the phenomena of violence and aggression, it is also necessary to present a classification of the phenomenon of aggression.

The classifications available in the literature distinguish at least a dozen types of aggression. It is usually based on various classification criteria that combine them into opposing pairs [Szydłowska, Jankowiak 2019, p. 335].

The first form mentioned is **psychological (verbal) aggression**, which includes abusive language, the use of profanity, insults, raised voices, shouting, screaming, threats, blackmail, intimidation, name-calling, mockery, disregard, humiliation, and unjustified criticism [Szydłowska, Jankowiak 2019, p. 335]. **Physical aggression**, on the other hand, is associated with a violation of physical integrity through direct assault, touching, beating, kicking, hitting, scratching, pinching, bumping, spitting, biting, holding down the body, twisting arms, pulling clothes or hair, pouring liquids, throwing objects, or attacks with tools [Grudzień et al. 2015, p. 243; Szydłowska, Jankowiak 2019, pp. 334–335; Florek, Siemiginowska 2020, p. 13].

The literature on the subject divides aggression into passive and active [Grudzień et al. 2015, p. 243]. **Active aggression** can take the form of verbal or physical aggression in the form of an act in which the attacker uses parts of the body or tools to inflict pain, suffering, or bodily harm on the attacked person. **Passive aggression**, on the other hand, is carried out through omission [Grudzień et al. 2015, p. 243]. The literature also presents other classifications of aggression, including direct and indirect aggression [Szydłowska, Jankowiak 2019, p. 339; Florek, Siemiginowska 2020, p. 14]. **Direct aggression** occurs when a person is aggressive toward an object or another person. An example of direct aggression is a physical attack (kicking, spitting, pulling, hitting in the face). A direct form of aggression can also be a verbal or psychological act (insults, mockery, raised voice, public humiliation of the victim). **Indirect aggression**, on the other hand, occurs when other people are used to carry out the aggression. It can also be a physical act (destroying the victim's property) or psychological (lying and spreading false information about the victim) [Tobiasz-Adamczyk 2014, p. 8; Szydłowska, Jankowiak 2019, p. 334; Florek, Siemiginowska 2020, p. 14]. In addition, a distinction is made between hostile (reactive) and instrumental (proactive) aggression [Lickiewicz 2017, p. 17; Florek, Siemiginowska 2020, p. 14]. **Hostile aggression** stems from the desire to harm someone and is very similar to physical aggression. Its purpose is to hurt or cause pain to another person [Lickiewicz 2017, p. 17; Florek, Siemiginowska 2020, p. 14]. According to Jacek Lickiewicz, **instrumental aggression** serves to achieve "some" goal, which may be to hurt or cause pain [Lickiewicz 2017, p. 17]. However, the main goal of instrumental aggression is to force submission and obedience [Fiłończuk-Wieczorkowska, Żukrowska 2014, p. 127]. Hostile aggression, on the other hand, is accompanied by the desire to achieve a specific goal and should be treated as a type of instrumental aggression [Florek, Siemiginowska 2020, p. 14].

The literature on the subject also distinguishes three other categories of aggression. The first is aggression in the form of manifesting hostility, the symptoms of which are raised voices, shouting, profanity, threats, blackmail, comments, offensive gestures, name-calling, insinuations, boycotts, excessive and unjustified criticism, false information, unfair evaluation of work, and ostentatious silence. The second category is hindering and preventing the performance of professional tasks, which is expressed by deliberately slowing down the work of the team. The third and most drastic category is overt aggression of a physical nature, which includes fatal physical attacks, murder, and attacks resulting in bodily injury such as pushing, beating, and shoving. Interpersonal (between people) aggression is significant from the point of view of healthcare work [Fiłończuk-Wieczorkowska, Żukrowska 2014, p. 128]. This type of aggression occurs in a relationship with another person, most often a patient or their family, but attacks in the form of physical aggression are also not uncommon [Ipek, Özlük 2023; Pallarés et al. 2023; Rabin et al. 2025; Tarczoń et al. 2025].

With the development of new technologies, a new type of aggression has emerged – **electronic aggression or cyberbullying.** It includes all types of aggression in which mobile phones and the Internet are the main tools of implementation [Sommer 2013, p. 283]. These may be actions directed directly against another person, e.g., insults via e-mail, or indirectly, e.g., posting information online that compromises a person. Cases of electronic aggression against medical personnel, especially doctors, are becoming increasingly common. Most often, they take the form of false and discrediting messages shared online [Tarczoń et al. 2025].

Some authors equate the concepts of violence and aggression because they have elements in common. Others believe that violence is related to aggression and constitutes "a type of aggression involving extreme forms of physical aggression" [Krahé 2005, p. 19; Tobiasz-Adamczyk 2014, p. 5]. Both verbal and physical aggression include behaviors of moderate intensity. Violence, on the other hand, refers to behaviors of the highest intensity [Dorosz, Kowalczuk 2019, p. 426]. Violence is a process, while aggression is a behavior [Czemierowska-Koruba 2015, p. 5]. The goal of aggression is to harm the victim, while the goal of violence is to influence them [Mazur 2002, p. 10]. Violence is a form of aggression, but not all forms of aggression are violence [Anderson, Bushman 2002, p. 29; Tobiasz-Adamczyk 2014, p. 5]. Pushing a child off a bicycle by another child is an act of aggression, but it is not an act of violence [Anderson, Bushman 2002, p. 29]. Violence is characterized by an imbalance of power between the perpetrator and the victim. The perpetrator always has the upper hand – the perpetrator is stronger than the victim [Gulla 2020, p. 15]. Aggression, on the other hand, concerns relationships between people with similar physical and mental predispositions [Czemierowska-Koruba 2015, pp. 5–6].

In summary, the concepts of violence and aggression are synonymous, but not identical, and their interchangeable use in certain cases is not acceptable, because in the narrow sense, violence is treated as equivalent to physical aggression [Rode 2010, p. 46]. In cases of aggression expressed in physical form, we can speak of violence, especially when acts of aggression are directed against women, who are by nature physically weaker and more delicate than men [Rode 2010, p. 46]. A more pragmatic term for patient behavior in healthcare facilities is aggression, due to the broader understanding of the concept, which includes all acts of violence. Therefore, it is assumed that **aggressive patient behavior is: a phenomenon that goes beyond the social rules of mutual relations, consisting in causing harm leading to physical and emotional consequences, poor mental and physical health of the employee, disruption of work, and even withdrawal from professional life, which causes significant costs for employees, their families, and society.**

A phenomenon that threatens the sense of security of employees in the workplace is violence, referred to in the literature as aggression or harassment. According to the International Labor Organization (ILO), **workplace violence** is defined as

> an act, incident, or behavior that deviates from reasonable cause, whereby a person is attacked, intimidated, humiliated, degraded, or injured by another person during or as a direct result of their work, or injured by another person during or as a direct result of their work.
>
> **[ILO, as cited in Pallarés et al., 2023, p. 353]**

The occurrence of violence in HC depends on the overlap of many **factors,** not only personal and interpersonal, organizational, but also social and cultural [Pallarés et al., 2023, p. 354]. These have consequences for key stakeholders, as they lead to many changes in the environment and adverse phenomena such as a decline in work performance and, consequently, in the quality of medical services provided, as well as a decline in job satisfaction.

There are two types of workplace violence based on its source: intra-organizational, when the perpetrator is another employee, and extra-organizational, when the perpetrator is a person from outside the organization. The causes of violence may lie in individual characteristics related to the personal situation (internal, individual) of the perpetrator and the victim, as well as in organizational factors (external, environmental) [Mościcka, Drabek 2010, p. 470]. Individual and organizational factors conducive to violence in the workplace have been collected and systematized in Table 4.3.

The literature on the subject indicates that the factors influencing the escalation of violence against healthcare workers are: the gender of the perpetrator (more often men), abuse of psychotropic substances, excessive alcohol consumption, mental illness, as well as high levels of hostility and anger [di Martino 2003, p. 17]. The appearance of the employee and the associated first impression in direct contact may be a factor encouraging or discouraging violence against medical staff. Another feature conducive to violence in healthcare facilities is the gender of the victim. The healthcare sector is a sector with a high level of female employment, including nurses. Nursing is a profession dominated by women in all countries. They account for over 80% of nursing staff in most countries around the world [di Martino 2003, p. 19]. The risk of violence against women in healthcare facilities is particularly high, especially in the case of a specific form of violence, namely sexual harassment. The mental and physical health of employees has a significant impact on their professional functioning. Health problems can lead to behaviors that affect interactions with patients and accelerate aggressive behaviors [di Martino 2003, p. 18]. Another victim characteristic that encourages or discourages violence is the age and experience of the employee. These characteristics are interrelated because previous experience in dealing with similar difficult situations is associated with age and enables these employees to respond better to similar incidents [di Martino 2003, pp. 18–19].

Many **organizational factors** can also contribute to workplace violence. One of them is inadequate working conditions – poor ventilation, poor lighting, temperature, noise, dirty rooms. Furthermore, improper work organization can lead to

TABLE 4.3
Factors Influencing Workplace Violence

Factors contributing to workplace violence		
Individual		
Perpetrator characteristics	**Victim characteristics**	**Organizational**
History of aggressive behavior	Physical appearance	Physical working conditions
Gender	Gender and sexual orientation	Workload (physical and mental) and pace of work
Age	Age and experience	Low wages
Experiencing difficulties in childhood due to inadequate parenting	Personality	No pay raises and limited career development opportunities (no promotion)
Difficult family relationships	Temperament	Assignment of the worst tasks
Poor academic performance	Attitudes and expectations	Ignoring ideas
Substance abuse problems	Health	Burdening with burdensome and unnecessary responsibilities
Excessive alcohol consumption	Skin color, race	Irregular working hours (shift work and night work)
Severe mental illness	Cultural or national identity	Communication problems
Being in situations conducive to violence	Problems with work efficiency	Lack of employee participation in safety
Access to firearms	Religious beliefs	Job insecurity

Source: Based on literature.

excessive workloads for a given group of employees and slow down their productivity, which in turn can cause negative attitudes among employees and trigger negative behavior in patients. Good lighting in the workplace and space to maneuver around patients and equipment are basic measures to reduce the risk of violence and aggressive behavior toward healthcare workers. Significant factors contributing to violence in the healthcare sector may include exposure to hazardous substances, the risk of infection with HIV, which causes AIDS, and infection with SARS-CoV-2, which causes COVID-19. Another factor contributing to workplace violence may be significant differences in pay for the same work and limited career development opportunities. Communication problems also contribute to workplace violence. Effective communication and proper information flow can relieve tension and frustration among employees and significantly prevent violence. The organization of working time plays a key role in the emergence of violence in healthcare. Working irregular hours causes a number of negative effects such as anxiety, fear, sleep disorders, digestive problems, and disorders related to changes in eating habits. An important factor driving violence in the workplace is restructuring, which is increasingly common in the healthcare sector and is associated with job insecurity and constant change. Uncertainty and fear of job loss affect commitment and social life. In organizations that are about to close or are undergoing restructuring, staff are less likely to be friendly and conciliatory in their dealings with clients or patients. Factors that can cause frustration and, as a consequence, contribute to the escalation of violence within an organization include assigning employees the worst tasks, ignoring ideas put forward by employees, burdening employees with onerous and unnecessary duties, and a lack of participation in safety measures. The relationship between the external environment and the work environment also seems to be important. In the context of possible violence in healthcare facilities, especially hospitals, workplace design requires special attention. Key requirements include public areas such as the size of waiting rooms, comfortable seating, availability of toilets, entrance supervision, alarms, escape routes, emergency plans, cameras, and surveillance systems that alert employees when urgent assistance is needed [di Martino 2003, pp. 20–23].

THE PHENOMENON OF PATIENT AGGRESSION IN THE HEALTHCARE SECTOR

Aggression toward healthcare personnel is a widespread and increasingly dangerous phenomenon. It is described as an "'epidemic" affecting staff recruitment, retention, and ability to provide high-quality care [Richardson et al. 2019, p. 271]. In addition, healthcare workers are the occupational group most exposed to aggression [Pallarés et al. 2023, p. 354] and harassment in the workplace [Wiskow et al. 2010, p. 15]. Surprisingly, the risk of exposure for medical personnel is even higher than for police officers and prison guards [Ziaei et al. 2019, p. 1]. Healthcare workers are eight times more likely to experience physical violence than manufacturing workers [Wiskow et al. 2010, p. 15]. According to the ILO, healthcare workers rank second in terms of exposure to violence in the workplace [Kowalczuk, Krajewska-Kułak 2017, p. 113]. Most of the incidents they experience are related to aggressive behavior by patients. This is confirmed by research by Daniela Grudzień and colleagues, which showed that staff encounter aggressive behavior by patients up to several times a week [Grudzień et al. 2015, p. 243].

Recent research conducted among NHS staff as part of The 2023 NHS Staff Survey indicates that aggressive patient behavior is a serious challenge that must be addressed. According to the results of the study:

- 13.7% of NHS staff have experienced at least one incident of physical violence from patients, service users, relatives, or other individuals in the last 12 months.
- 27.6% of ambulance staff reported experiencing at least one incident of physical violence from patients, service users, relatives, or others in the last 12 months.
- 25.15% of NHS staff experienced at least one incident of harassment, intimidation, or abuse in the last 12 months from patients/service users, their relatives, or other individuals, which is lower than in previous years.

Aggressive behavior takes different forms and has different intensities in outpatient clinics, pediatric wards, psychiatric wards, and emergency departments.

[Kowalczuk, Krajewska-Kułak 2017, p. 115]. However, within the HC sector, it is emergency department (ED) staff who are most exposed to various forms of violence [Pallarés et al. 2023, p. 354]. The scale of this phenomenon (violence in EDs in HC) varies depending on the geographical location of a given country. And despite the lack of accurate estimates, the available data from individual countries show alarming statistics. According to research, in Germany as many as 87% of ED employees have experienced physical violence and 97% have experienced verbal aggression [Rabin et al., 2025]. In the United Kingdom, 96% of respondents were victims of intimidation or assault [Rabin et al. 2025], in the US 71% [Rabin et al., 2025], in Italy 64%, and in Spain it is estimated that 60% of ED workers have experienced violence in the workplace [Pallarés et al. 2023, p. 354]. Unfortunately, these figures are usually underestimated [Ipek, Özlük 2023], and in many countries such data is lacking.

Studies have shown that nursing staff, midwives, and doctors are all vulnerable to aggressive behavior in the workplace, but most studies refer to nursing staff. More than two-thirds of doctors have experienced work-related assaults, but among all professions, nurses are the most exposed to aggressive behavior [Ramacciati et al. 2016]. According to studies conducted in various parts of the world, between 59.2% and 94.6% of nurses surveyed have experienced violence in the workplace, experiencing various forms of violence, ranging from verbal aggression to sexual harassment [Ramacciati et al. 2016].

The risk factors for aggressive behavior on the part of patients vary. The literature points to variables such as the patient's gender and age. Studies available in the literature have shown that aggressive behavior is more common in men aged 20–29 or 60 and older [Lickiewicz 2017, p. 19]. However, the results of the studies are not consistent in this respect, but age certainly has a significant impact on aggressive verbal behavior [Grudzień et al. 2015, p. 243], which is more intense among young people and weakens with age [Szkup et al. 2018, p. 121]. This is also confirmed by research conducted on a group of emergency medical services employees [Frydrysiak et al. 2014, pp. 381–391; Mikos et al. 2018, pp. 57–65]. However, there is a group of respondents who believe that gender does not significantly affect the risk of aggressive behavior [Grudzień et al. 2015, p. 243]. Thus, research in this area is inconclusive.

The most common **causes of aggressive behavior toward healthcare workers** include head injuries, mental retardation, mental disorders (schizophrenia), personality disorders (hallucinations, paranoid delusions, excessive suspiciousness), and the treatment environment [Lickiewicz 2017, p. 19]. Mental illness, craniocerebral injuries, alcohol intoxication, and addiction to psychoactive substances are indicated as causes of aggressive behavior in patients [Grudzień et al. 2015, p. 244]. Available literature studies indicate that the causes of aggressive behavior in patients also include: lack of acceptance by the environment, past sexual abuse, problems with the law, domestic violence, negative assessment of the actions of medical staff, inappropriate treatment of patients and their families, and lack of understanding of the patient's situation [Frydrysiak et al. 2014, pp. 382–388]. Wearing clothing assigned to members of a particular professional group may also encourage aggressive behavior [di Martino 2003, p. 19]. According to the World Health Organization, aggressive behavior in patients can be caused by anxiety, fear of danger, cognitive disorders, narrowing of consciousness, which make it difficult to assess the situation, as well as other diseases [WHO & ILO (World Health Organization and International Labour Organization) 2018, p. 46]. The source of aggressive patient behavior may also be media reports and the functioning of a given medical facility [Kmiecik 2016, pp. 159–161]. One of the elements of the functioning of a medical facility that can cause aggressive behavior in patients is long waiting times for medical advice [Szwamel, Sochocka 2014, p. 153; di Martino 2003, p. 21]. The consequence of long waiting times is patient frustration, which triggers aggressive behavior toward medical staff. Frustration also stems from patients' insufficient knowledge of how the ED works. The order in which medical advice is given in the ED is not based on the order in which patients arrive at the facility, but depends on the patient's health as determined by medical staff during **the *triage* procedure**.[1] Often, patients and their loved ones are unaware that patients in life-threatening conditions have priority in the ED.

Thus, the occurrence of violence in HC depends on the overlap of many factors, not only personal and interpersonal, organizational, but also social and cultural [Pallarés et al., 2023, p. 354]. These factors have consequences for key stakeholders, as they lead to the deterioration of employees' health and many changes in the environment, as well as adverse phenomena such as a decline in work performance and, consequently, in the quality of medical services provided, or a decrease in job satisfaction. When analyzing the effects, it should be mentioned that there are even cases of violence resulting in the death of a healthcare worker. For example, in Turkey, between 2005 and 2022, as many as 11 cases of death after a knife attack on healthcare workers were reported [Ipek, Özlük 2023]. The most common perpetrators of this violence are the patients themselves who are receiving medical assistance, but also their families [ALBashtawy, Aljezawi 2016; Yang, Stone, Petrini et al. 2018].

An analysis of the literature indicates that the phenomenon of aggressive patient behavior is a growing problem [Grudzień et al. 2015, Ipek, Özlük 2023, Pallarés et al., 2023; Rabin et al., 2025; Tarczoń et al., 2025]. Therefore, it is important to conduct an appropriate information campaign among the public, unequivocally condemning violence against medical staff and pointing out the consequences of such behavior, as well as publicizing not only the rights of patients but also their obligations.

THE PATIENT'S PERSPECTIVE AND THE PHENOMENON OF AGGRESSION IN THE HEALTHCARE SECTOR

One of the first steps toward greater public awareness of emerging violence in the healthcare sector is to conduct public opinion polls or surveys of patients themselves on this topic. There are still very few such studies [Ipek, Özlük 2023; Thanh Tùng et al. 2021]. Studies conducted with patients as respondents indicate that one in four patients admitted to hospital is sometimes aggressive toward medical staff, mainly nurses [Ipek, Özlük 2023]. However, most of the cases of violence reported by patients involved verbal or psychological acts of aggressive behavior.

In the age of social media, various incidents are increasingly being commented on, including acts of aggressive behavior by patients toward medical staff. Some researchers have analyzed patients' reactions to violence against medical staff. What do patients themselves think about this issue?

Research conducted in Vietnam, which analyzed readers' comments in response to published cases of patient violence against medical staff, revealed the patients' perspective [Thanh Tùng et al. 2021]. A thematic analysis of the published comments identified four main attitudes among patients:

- proposal of punishment, i.e., readers' outrage and desire for harsher punishment of perpetrators of such acts;
- understanding and sympathy for the injured medical staff;
- a skeptical approach to such incidents, where patients who had previously experienced poor-quality medical services wondered about the perpetrator's motives;
- seeing the conflict more between people of different genders (women and men) than between the patient and the nurse, as in this case the commentators focused on nurses as women.

Although patients' opinions vary, looking at their scale, those condemning aggressive behavior predominate. Patients themselves expect more severe penalties for attacking patients and aggressive family members, as in some countries the penalties are inadequate to the harm caused, e.g., in Vietnam or Turkey [Thanh Tùng et al. 2021]. According to the results of previous studies, patients expect better measurement and recording of such incidents in order to be able to counteract them effectively. Surprisingly, patients note that cases that may cast doubt on the intentions or working methods of medical staff, thereby putting them in a bad light, should not be made public and publicized on social media [Ipek, Özlük 2023]. This attitude indicates that patients identify with medical staff and seek realistic ways to counteract such incidents. Moreover, the respondents indicate that it is necessary to change the prevailing opinion that violence in the workplace is, in a sense, an integral part of the work of medical staff. Similarly, it seems necessary not only to introduce processes and tools to measure this phenomenon, but above all to change the perception of medical staff themselves regarding the reporting of incidents, which is necessary to estimate the scale of the phenomenon and help take steps to de-escalate it [Ipek, Özlük 2023].

When discussing the opinions of patients themselves, it is also worth mentioning the public perception of the use of weapon detection systems in EDs in urban areas, which has been the subject of previous studies. The results of these studies indicated that two-thirds of respondents believe that such systems should be in place in EDs, with 71% of those in favor of weapon detection systems being women [McNamara, Yu, Kelly 1997].

The perception of the patients themselves indicates increasing awareness and great solidarity with medical staff who experience violence. Patients are looking for ways to de-escalate the problem, pointing to the need for accurate records, the imposition of severe penalties for both patients and families who commit violence against medical staff, and the introduction of specific tools to build a sense of security in the EDs.

SUMMARY

Violence and aggression in the workplace pose a serious threat to employees, causing many negative consequences. Healthcare workers rank second in terms of exposure to violence in the workplace [Kowalczuk, Krajewska-Kułak 2017, p. 113]. The problem of the growing number of cases of aggressive behavior by patients toward healthcare workers has been particularly noticeable during the ongoing SARS-CoV-2 epidemic [Health Policy.com, 2021].

Due to the different understandings of the concepts of violence and aggression, this book uses the term "**aggressive patient behavior**," which is defined as behavior that goes beyond the social rules of mutual relations and involves causing harm leading to physical and emotional consequences, as well as poor mental and physical health. According to research, aggressive behavior toward medical staff is most often committed by men, and the reasons for such behavior are primarily mental illness, traumatic brain injury, alcohol intoxication, and addiction to psychoactive substances. Unfortunately, this problem may be exacerbated by one-sided media coverage and the focus on patients' rights, as well as by changing behavior patterns (increasingly demanding attitudes) and the greater sense of threat and stress that we feel as a society.

Previous studies often point to the powerlessness of medical staff and their isolation in combating this phenomenon, despite the high costs incurred by employees, their families, employers, and even society. Some countries lack adequate legal provisions and even organizational procedures to effectively counteract or support employees in combating aggressive patient behavior, even in developed countries. The problem of aggressive patient behavior toward medical staff is also widely commented on by patients themselves, who often sympathize with the victims of such violence and demand harsher penalties for them. In view of the above, efforts are needed to change this situation and improve the working conditions of medical staff in such a way as to build a more harmonious working environment that supports a high quality of professional life (well-being) for this group of employees.

Research indicates that emergency department staff are among the most vulnerable professional groups in this regard. It is difficult to accurately determine the scale of this problem in some countries, as there is no central register of such incidents. Therefore, the presented research results do not fully reflect the extent of aggressive behavior by patients toward medical staff. Most of the studies presented concern only nursing staff. However, there is a lack of detailed research on aggressive behavior of patients toward doctors and paramedics.

NOTE

1 Triage is a term derived from French and refers to emergency medical triage, which allows determining which patients need immediate medical attention and which can be treated later. In Poland, the "START" triage system is widely used, which stands for *Simple Triage and Rapid Treatment*.

5 Behind the Triage
Research on Patient Aggression in Emergency Departments

In order to show the scale of aggressive behavior by patients toward emergency department (ED) medical staff and the most common forms of aggressive behavior by patients, a survey was conducted among doctors, nursing staff, and paramedics in EDs in Poland. The choice of country is dictated by the fact that the scale of this phenomenon in Poland is unknown, and many such incidents are not reported.

The empirical research consisted of two stages. The first stage of empirical research consisted of surveys conducted among ED staff in Poland, which allowed to learn about the scale and specifics of the phenomenon of aggressive patient behavior.

The second stage consisted of interviews with experts (semi-structured interviews), which allowed for a deeper understanding of the essence of the problem, and in particular, the consequences of aggressive patient behavior for individual stakeholders and the reasons for the lack of response or reporting of cases of aggression by medical personnel. The results of this stage of the research led to further qualitative research – an analysis of cases of aggressive patient behavior that had particularly severe consequences for medical staff. The research procedure is briefly illustrated in Figure 5.1.

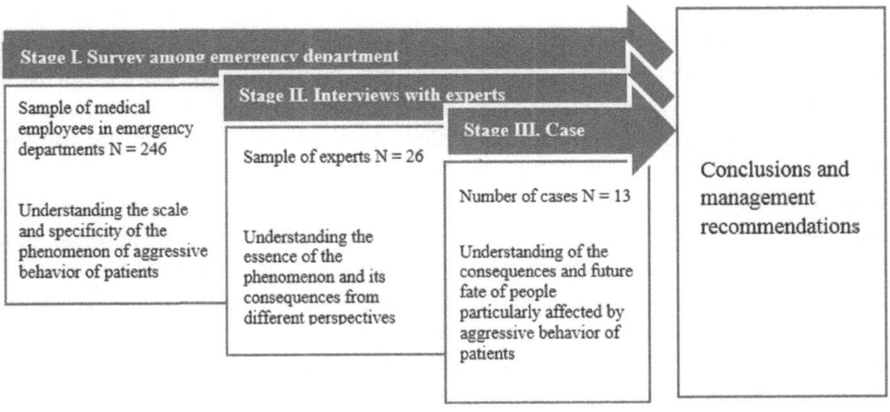

FIGURE 5.1 Research procedure diagram

DOI: 10.1201/9781003716631-6

SURVEY AMONG EMERGENCY DEPARTMENT EMPLOYEES

Due to the subject of the study, a purposive sample was selected for the study – medical facilities with EDs in their structures. Requests for consent to conduct the survey were sent to the management of the medical facilities. Unfortunately, due to the ongoing epidemic and the need to take measures to limit the spread of the virus, direct contact with respondents was not possible. The ongoing epidemic of SARS-CoV-2 infections and the rate at which it is spreading have caused a shortage of medical personnel. Therefore, requests were sent to the management of randomly selected medical facilities in Poland asking for their consent to conduct the survey. A total of 26 healthcare entities with hospital EDs were invited to participate in the study. The number of selected EDs accounted for 11.0% of all EDs operating in Poland. The study was conducted from September to December 2020 in a manner that ensured the anonymity of the participants. The questionnaires were completed by doctors, nursing staff, and paramedics working in EDs in Poland. As a result of the survey, 249 questionnaires were obtained, of which three were rejected because they indicated that the respondent had no experience in dealing with aggressive patients.

Research Tool

The questions in the survey questionnaire were divided into four thematic sections. Section I concerned the experiences of medical personnel with aggressive patient behavior. The questions in this section referred to incidents of aggressive patient behavior, their type, and frequency. Section II covered issues related to the causes and consequences of aggressive behavior. The questions in this section concerned the factors generating aggressive behavior in patients and their companions, the consequences of such behavior, and how to deal with patients who behave aggressively. Section III concerned the organization of work and management in hospital EDs. This section of the survey consisted of questions about the impact of aggressive behavior on the organization's activities. The questionnaire contained 11 questions (12–22) using a five-point Likert scale. Each answer was assigned a numerical value from 1 to 5 points (1 – strongly disagree, 2 – disagree, 3 – no opinion, 4 – agree, and 5 – completely agree), which allowed for obtaining numerical information and determining which answers were selected most often in a given group of respondents. This part of the questionnaire also included questions about the need for training in dealing with patients who behave aggressively toward medical staff. Section IV contained personal details, which included the following demographic data about the respondent: gender, profession (medical), length of service, housing situation (whether the respondent lives alone or with other people), number of people living together in the household, size of the organization where the respondent works (expressed as the number of employees), form of employment (type of contract), size of the city, and province in which the medical facility is located. It was important to obtain an answer to the question of whether the respondent lives with other people due to the possibility of obtaining support in stressful situations at work, such as aggressive patient behavior.

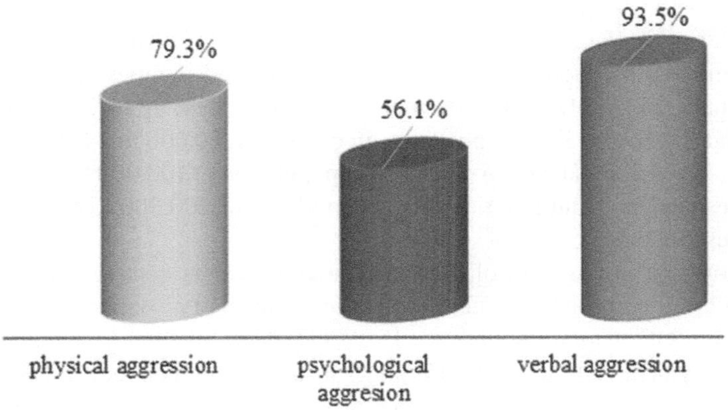

FIGURE 5.2 Structure of respondents by gender and profession

Research Sample

A total of 249 respondents took part in the questionnaire survey. Ultimately, data obtained from 246 respondents were analyzed – 58.9% (145) women and 41.1% (101) men. All respondents participating in the survey were employees of hospital EDs.

The group of physicians included 63 respondents – 44.4% (28) women and 55.6% (35) men. The nursing staff included 89 respondents – 95.5% (85) women and 4.5% (4) men. The group of paramedics included 94 respondents – 34.0% (32) women and 66.0% (62) men (Figure 5.2).

The majority of respondents were employees with short work experience of 1 to 5 years. The smallest group consisted of respondents with long work experience – over 35 years.

Most respondents work in large medical facilities that employ over 1,000 people – 37.8% (93), and the fewest in hospitals that employ up to 500 people – 13.0% (32). The characteristics of respondents by length of service and size of organization are presented in Table 5.1.

TABLE 5.1
Structure of Respondents by Length of Service and Size of Organization

	Organization Size [Number of Employees]							
	Up to 500		500–750		750–1,000		>1,000	
Length of Service	N	[%]	N	[%]	N	[%]	N	[%]
1–5 years	10	31.3	12	29.3	25	39.7	31	33.3
6–15 years	6	18.8	2	4.9	10	15.9	12	12.9
16–25 years	9	28.1	10	24.4	13	20.6	26	28.0
26–35 years	3	9.4	14	34.1	11	17.5	16	17.2
>35 years	4	12.5	3	7.3	4	6.3	8	8.6

Source: Based on research results.

The largest group consisted of residents of large cities with over 500,000 inhabitants – 37.8% (94) – and rural residents – 19.7% (49). The surveyed group was also dominated by people employed in entities located in cities with over 500,000 inhabitants – 51.2% (126). Those employed in cities with up to 20,000 inhabitants accounted for 9.4% (23) of respondents, in cities with 20,000 to 100,000 inhabitants – 18.7% (46) of respondents, in cities with a population of 100,000 to 200,000 – 8.5% (21) of respondents, and in cities with a population of 200,000 to 500,000 – 12.2% (34) of respondents.

Employment in the form of a permanent employment contract was declared by 50.8% (125) of respondents, and a contract in the form of a contract by 32.1% (79) of respondents. Of the total respondents, 15.0% (37) declared employment under a fixed-term employment contract, 2.0% (5) under a substitution contract, 1.2% (3) under a trial period contract, and 2.4% (6) under a contract of mandate. The answers do not add up to 100% because respondents gave more than one answer, providing work on the basis of an employment contract and on the basis of a contract agreement.

Analysis of Survey Results

The collected data show that ED employees experience various forms of aggressive behavior from patients. Physical forms of aggressive behavior were experienced by 79.3% of respondents, psychological forms by 56.1% of respondents, and verbal forms by 93.5% of respondents (Figure 5.3). For greater clarity in the tables and figures, physical, psychological, and verbal forms of aggressive behavior by patients will be described as physical aggression, psychological aggression, and verbal aggression.

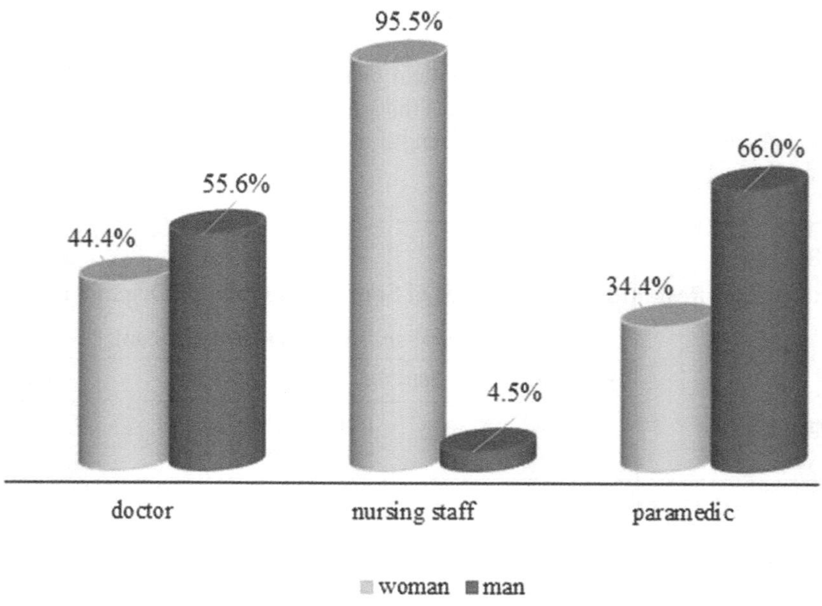

FIGURE 5.3 Forms of aggressive behavior in patients

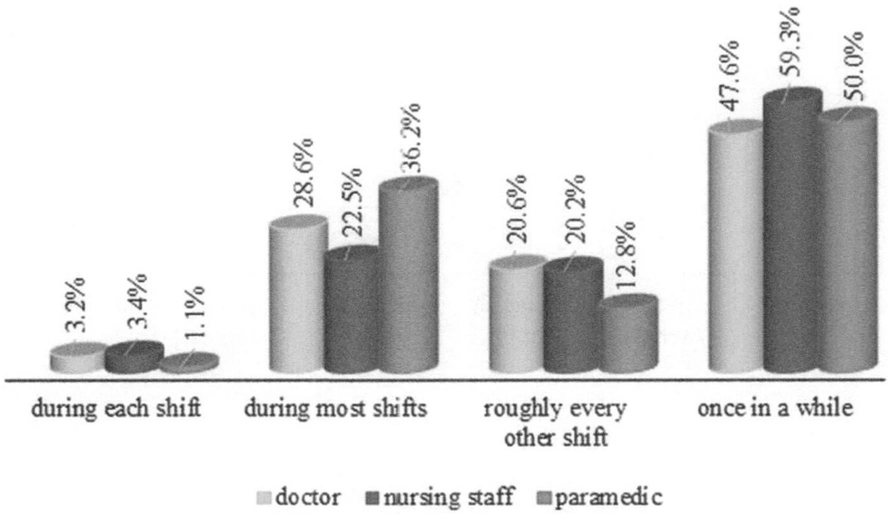

FIGURE 5.4 Frequency of experiencing aggressive behavior from patients

Over 50% of the surveyed ED staff experienced aggressive behavior from patients "once in a while." In turn, 23.9% of ED staff experienced aggressive behavior from patients "during most shifts," 17.5% experienced aggressive behavior from patients "roughly every other shift," and 2.4% experienced aggressive behavior from patients "during every shift." The highest percentage of ED medical staff who experienced aggressive behavior from patients "once in a while" was observed among nursing staff. Among the respondents who experienced aggressive behavior from patients "during most shifts," the highest percentage was observed in the group of paramedics. No significant differences were observed among doctors and nursing staff who experienced aggressive behavior from patients with a frequency of "approximately every other shift" and "during every shift." A small percentage of doctors, nursing staff, and paramedics experienced aggressive behavior from patients "during every shift" (Figure 5.4).

Due to the 24-hour nature of ED work, it was important to examine at what time of day patients most often behave aggressively. In the survey, 51.6% of respondents did not specify at what time patients behaved aggressively. Of those surveyed, 31.7% believed that patients most often behaved aggressively at night, 22.4% believed that it was in the evening, 0.8% believed that it was in the morning, 3.3% believed that it was during the day, and 1.6% believed that it was in the middle of the day. Finally, 51.6% did not find any correlation here, indicating the answer "difficult to say" (Figure 5.5).

The percentage of women (79.3%) and men (79.2%) who experienced aggressive behavior from patients in physical form did not differ. In the case of psychological aggression, a slightly higher percentage of respondents who experienced this form of aggressive behavior was observed in the group of men (58.4%) than in the group of women (54.5%). In terms of experiences of verbal aggression, a slightly higher

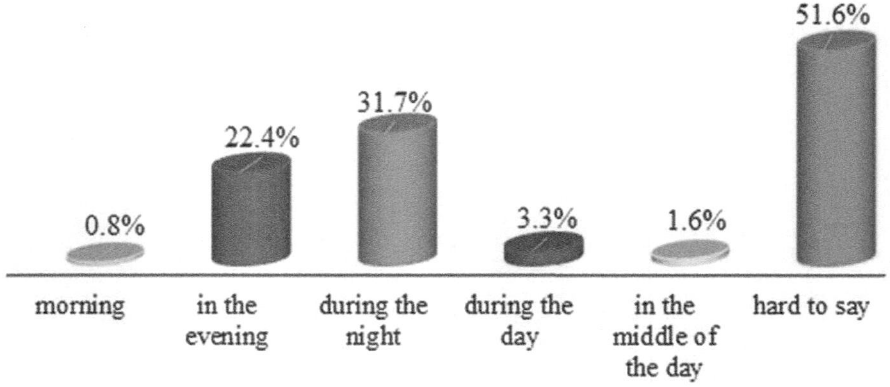

FIGURE 5.5 Aggressive behavior of patients depending on the time of day

percentage was also observed in the group of men (95.0%) than in the group of women (92.4%) (Figure 5.6). In response to the question of which gender is more exposed to aggressive behavior from patients, more than half of the respondents (55.3%) indicated women.

The research indicates that the most common physical forms of aggressive behavior by patients were waving their arms, pulling, pushing or shoving, and hitting or attempting to beat someone up. Respondents also experienced aggressive behavior in the form of kicking, biting, and hair pulling, which occurred much less frequently. There were also occasional cases of choking (Table 5.2).

When analyzing the frequency of physical forms of aggressive behavior exhibited by patients, the results show that respondents experience them "once in a while." The most common forms of physical aggressive behavior encountered by employees were hitting/attempted beating, pulling, pushing/shoving, and kicking (Table 5.3).

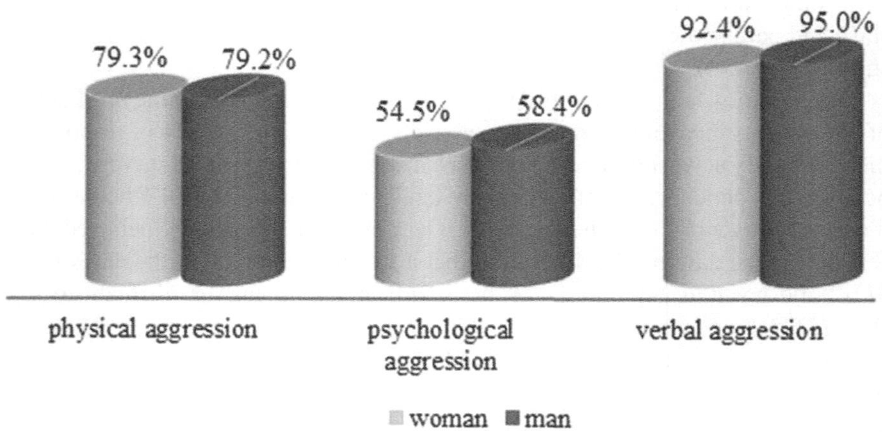

FIGURE 5.6 Forms of aggressive behavior by patients and gender of ED medical staff

TABLE 5.2
Physical Forms of Aggressive Behavior According to Frequency of Occurrence among Respondents

Physical Form of Aggression	N	[%]
Arm swinging	209	85.0
Jerking	166	67.5
Hitting/attempted assault	130	52.8
Kicking	118	48.0
Biting	81	32.9
Hair pulling	37	15.0
Suffocation	18	7.3
Push/push away	16	61.0

Source: Based on research results.

TABLE 5.3
Frequency of Physical Forms of Aggressive Behavior by Patients

	During Each Shift		During Most Shifts		Approximately Every Other Shift		Once in a While		Never	
Physical Aggression	N	%	N	%	N	%	N	%	N	%
Jerking	12	5.9	20	9.8	23	11.3	111	54.4	37	18.2
Hitting/attempted assault	3	1.6	15	8.2	11	6.0	101	54.9	53	29.0
Kicking	1	0.5	11	6.0	17	9.3	89	48.6	64	35.2
Biting	0	0	3	1.8	14	8.3	64	38.1	86	51.5
Push/push away	4	2.1	20	10.6	24	12.8	102	54.3	37	19.8
Arm swinging	21	9.5	54	24.5	47	21.4	88	40.0	9	4.1
Hair pulling	1	0.6	1	0.6	6	3.7	30	18.6	122	76.7
Suffocation	0	0.0	0	0.0	0	0.0	18	11.2	143	88.8

Source: Based on research results.

The most common psychological forms of aggressive behavior were intimidation, threats, humiliation, blackmail, ridicule, and slander. Persistent harassment (reporting and written complaints) was less common (Table 5.4).

The survey asked respondents about the frequency of psychological forms of aggressive behavior experienced from patients (Table 5.5). According to the data collected, respondents experience psychological forms of aggressive behavior from patients "once in a while." The most common psychological form of aggressive behavior encountered by respondents was false public accusation or slander, threats, and intimidation. Almost half of the respondents experienced blackmail, ridicule,

TABLE 5.4
Psychological Forms of Aggressive Behavior in Patients

Psychological Form of Aggression	N	%
Intimidation	180	73
Threat	177	72.0
Humiliation	151	61.4
Mocking	140	56.9
False public accusation/slander	138	56.1
Persistent harassment (denunciation, written complaint)	74	30.1
Blackmail	15	58.5

Source: Based on research results.

TABLE 5.5
Frequency of Psychological Forms of Aggressive Behavior by Patients

Psychological Aggression	During Every Shift		During Most Shifts		Approximately Every Other Shift		Once in a While		Never	
	N	%	N	%	N	%	N	%	N	%
Blackmail	9	5.1	25	14.1	25	14.1	85	48.0	32	18.2
Threat	14	7.5	27	14.4	37	19.8	99	52.9	9	4.8
Intimidation	10	5.2	32	16.5	40	20.6	98	50.5	13	6.7
Mocking	10	5.8	22	12.8	24	14.0	84	48.8	31	18.1
Humiliation	8	4.4	18	9.9	38	21.0	87	48.1	29	16.1
False public accusation, slander	7	3.9	18	10.1	13	7.3	100	55.9	40	22.5
Persistent harassment (reporting, written complaint)	4	2.4	5	3.0	9	5.4	56	33.3	93	55.7

Source: Based on research results.

and humiliation "once in a while" in their professional work. However, more than half have never experienced persistent harassment from patients.

At the same time, a survey was conducted among emergency room medical personnel who had experienced verbal forms of aggressive behavior from patients. Most of the respondents had experienced aggressive behavior in the form of shouting, swearing, verbal abuse, and verbal altercations. An equally high percentage of respondents had encountered insults directed at staff, defamation, and verbal abuse (Figure 5.7).

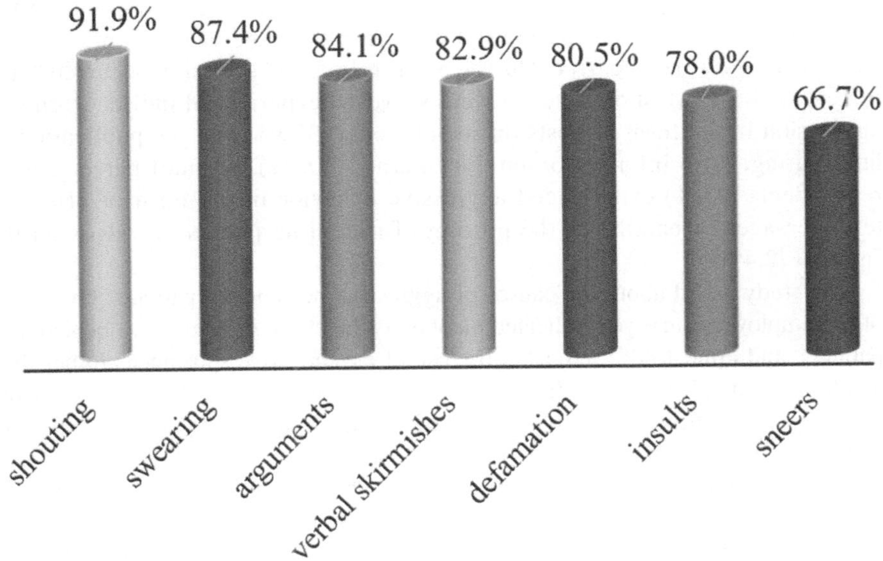

FIGURE 5.7 Verbal forms of aggressive behavior

The questionnaire also included a question about the frequency of verbal forms of aggressive behavior toward ED staff (Table 5.6).

Verbal aggressive behavior by patients was the most common. About 40% of respondents were "occasionally" subjected to verbal abuse, insults, or name-calling. Over 30% of respondents "occasionally" encountered aggressive behavior by patients in the form of shouting, verbal altercations, verbal abuse, and

TABLE 5.6
Frequency of Verbal Forms of Aggressive Behavior

	During Every Shift		During Most Shifts		Approximately Every Other Shift		Once in a While		Never	
Verbal Aggression	N	%	N	%	N	%	N	%	N	%
Shouting	49	21.2	66	28.6	36	15.6	75	32.5	4	2.2
Verbal skirmishes	32	15.2	63	29.9	37	17.5	72	34.1	5	3.3
Insults	20	9.8	48	23.4	41	20.0	83	41.0	11	5.9
Name-calling	18	8.5	48	22.7	39	18.5	93	44.1	12	6.2
Sneers	11	5.8	29	15.2	42	22.0	82	42.9	26	14.1
Arguments	44	20.5	45	20.9	46	21.4	72	33.5	7	3.7
Curses	67	30.2	45	20.3	32	14.4	71	32.4	5	2.7

Source: Based on research results.

swearing. In the opinion of almost 30% of respondents, shouting and verbal altercations occurred "during most shifts." Insults, name-calling, verbal abuse were also forms that occurred "during most shifts," and swearing occurred "during every shift." Most of the employees surveyed experienced indirect forms of aggression in the form of posts on social media (37.8%) and the publication of humiliating, false information on the Internet (20.3%). A small percentage of respondents (4.9%) experienced aggressive behavior in the form of offensive text messages or emails, or the posting of ridiculing photos or videos on the Internet (2.4%).

The study asked about the causes of aggressive behavior in patients. Over 90% of the employees surveyed indicated alcohol as the cause of aggressive behavior in patients, and almost 80% indicated the use of psychoactive substances (other than alcohol). Respondents also indicated negative perceptions of staff actions, pain and suffering due to illness, difficult life situations, and a lack of security as causes of aggressive behavior in patients.

The respondents' answers to the question about their reaction to aggressive behavior of patients in the ED indicate that the most common reaction of medical staff was to ask for a change in behavior. More than half of the respondents asked their colleagues and the police for help in cases of aggressive physical and psychological behavior. Approximately 40% of respondents ignored all forms of aggressive behavior by patients (Table 5.7).

As a result of contact with patients exhibiting all forms of aggressive behavior, emergency room medical personnel most often experienced anger, agitation, helplessness, and rage. Sadness, regret, and depression were less common (Table 5.8). The frequent occurrence of helplessness is worrying as it may indicate a lack of formal tools to support medical staff in dealing with incidents of violence.

TABLE 5.7
Staff Reactions to Aggressive Patient Behavior

	Aggression					
	Physical		Mental		Verbal	
Staff Reaction	N	[%]	N	[%]	N	[%]
Ignoring behavior	75	38.5	57	41.3	87	37.8
Discussion with the patient	51	26.2	39	28.3	57	24.8
Request for change in behavior	148	75.9	105	76.1	179	77.8
Use of sedative treatment	68	34.9	50	36.2	71	30.9
Forced immobilization	70	35.9	53	38.4	73	31.7
Seeking help from the police	101	51.8	73	52.9	113	49.1
Asking colleagues for help	109	55.9	85	61.6	115	50.0

Source: Based on research results.

TABLE 5.8
Emotions Experienced by ED Staff after Contact with a Patient Exhibiting Various Forms of Aggressive Behavior

	Aggression					
	Physical		Mental		Verbal	
Emotions	N	[%]	N	[%]	N	[%]
Anger	87	44.6	66	47.8	104	45.2
Anxiety and fear	72	35.9	56	40.6	80	34.8
Helplessness	78	40	56	40.6	94	40.9
Uncertainty	55	28.2	42	30.4	65	28.3
Anger	67	34.4	59	42.8	79	34.3
Sadness	27	13.8	22	15.9	31	13.5
Regret	27	13.8	19	13.8	27	11.7
Collapse	23	11.8	20	14.5	27	11.7
Agitation	80	41.0	57	41.3	96	41.7

Source: Based on research results.

The next questions concerned the perception of medical staff of selected aspects related to aggressive behavior of patients in the ED (Table 5.9).

The respondents' answers show the extent to which they agree or disagree with the statement in question. The answers show that, in the opinion of medical staff, aggressive patient behavior is an inherent part of working in the emergency room. Almost half of the respondents agree with the view that patients behave aggressively because they have to wait too long for medical advice. Most of the employees surveyed disagree with the statement that patients behave aggressively because they need to be hospitalized, or with the statement that aggressive behavior by patients during their stay in the ED is justified.

The surveyed employees disagree with the statement that the employer is interested in information about aggressive behavior of patients in the emergency room or with the statement that incidents of aggressive behavior of patients during their stay in the emergency room are not recorded in any documentation.

This may mean that despite the existence of appropriate registers or procedures for reporting incidents of aggressive patient behavior, they are ineffective. It is possible that this is related to the perceived "lack of interest" in this issue on the part of the employers of the respondents surveyed. The responses also suggest a need to raise awareness of the negative consequences for employers resulting from such incidents, which may translate into greater commitment to counteracting such events.

When analyzing the causes of incidents on the part of medical staff, the respondents indicate that the reason for the aggressive behavior of patients could be insufficient time devoted by staff to explaining the situation and procedures (42.3%), provocative behavior on the part of medical staff (22.1%), negligence in treatment

TABLE 5.9
Employees' Perspective on Aspects of Work Organization Related to Aggressive Patient Behavior

		Answers									
NR	**Question**	**1**	**%**	**2**	**%**	**3**	**%**	**4**	**%**	**5**	**%**
12	Aggressive patient behavior is an inherent part of working in the emergency room.	45	18.3	32	13.0	35	14.2	61	24.8	73	29.7
13	The patient behaves this way due to limited access to medical services.	39	15.9	54	22.0	68	27.6	60	24.4	25	10.2
14	The patient behaves aggressively because he has been waiting too long for medical advice.	18	7.3	47	19.1	61	24.8	76	30.9	44	17.9
15	The patient behaves aggressively due to dissatisfaction with the range of medical services offered.	30	12.2	47	19.1	82	33.3	57	23.2	30	12.2
16	The patient often behaves aggressively when medical procedures are performed.	41	16.7	56	22.8	75	30.5	57	23.2	17	6.9
17	The patient behaves aggressively due to the refusal to issue a certificate of incapacity for work, a medical certificate, a prescription, a referral for tests, etc.	32	13.0	59	24	73	29.7	60	24.4	22	8.9
18	The patient behaves aggressively due to the need for inpatient treatment – hospitalization.	72	29.3	65	26.4	67	27.2	29	11.8	13	5.3
19	In the ward where I work, the employee is blamed for the patient's aggressive behavior.	54	22.0	66	26.8	67	27.2	37	15.0	22	8.9
20	Aggressive behavior of the patient during their stay in the emergency room is justified.	89	36.2	67	27.2	49	19.9	23	9.3	18	7.3
21	Incidents of aggressive patient behavior during their stay in the emergency department are not usually recorded in any documentation.	58	23.6	67	27.2	60	24.4	40	16.3	21	8.5
22	The employer is interested in information about aggressive behavior of patients in the emergency room.	76	30.9	35	14.2	59	24.0	44	17.9	32	13.0

Source: Based on survey results.

Legend: 1 – "I completely disagree," 2 – "I disagree," 3 – "It's hard to say," 4 – "I agree," 5 – "I completely agree."

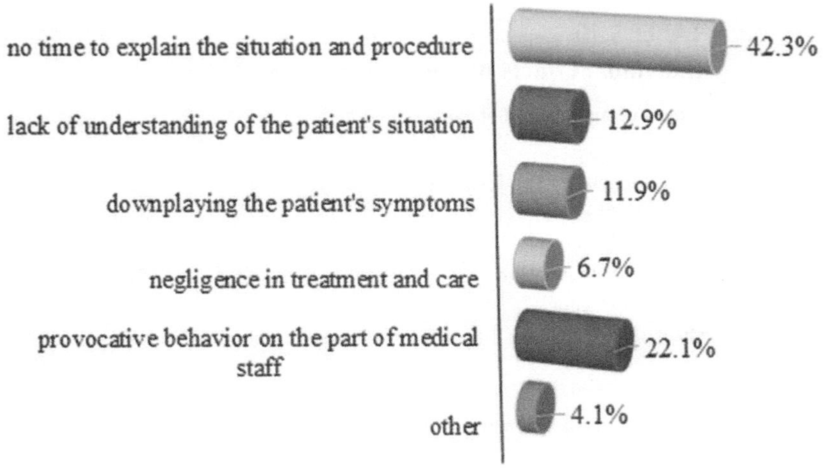

FIGURE 5.8 Staff behavior and aggressive patient behavior

and care (6.7%), lack of understanding of the patient's situation (12.9%), and downplaying of the patient's complaints (11.9%). In the "other" category (4.1%), respondents indicated contemptuous treatment of staff and procedures that do not meet patient expectations (Figure 5.8).

More than half of the respondents believed that patients are more likely to behave aggressively in the presence of accompanying persons: family members, partners, as well as other patients and security personnel. Among the persons accompanying the patient, the majority of respondents indicated the patient's family and friends as the perpetrators of aggressive behavior (Figure 5.9).

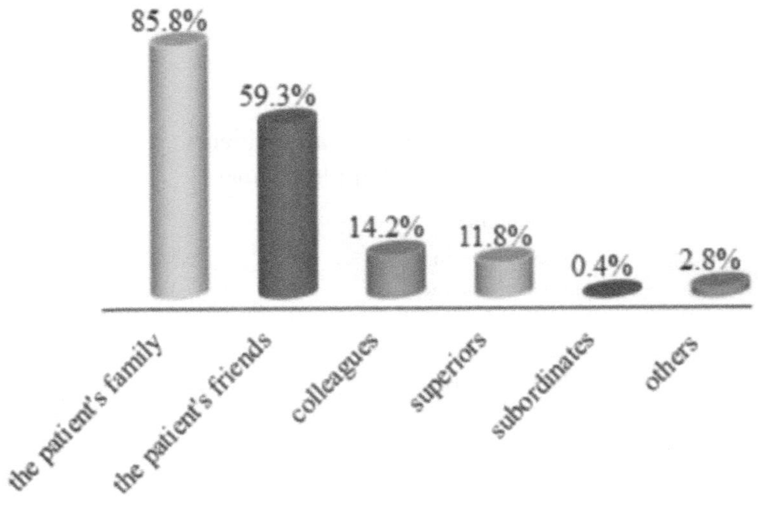

FIGURE 5.9 Accompanying persons as perpetrators of aggressive behavior

TABLE 5.10
Individual Effects of Aggressive Patient Behavior

Effects of Aggressive Behavior	N	%
Long-term stress	65	26.4
Loss of dignity and professional pride	72	29.3
Low self-esteem	47	19.1
Depressive states	16	6.5
General health problems	68	27.6
Job dissatisfaction	123	50.0
Other	24	9.8

Source: Based on survey results.

In response to the question about the perceived individual effects of aggressive behavior by patients, most respondents, as many as 50%, indicated dissatisfaction with their work, followed by a loss of dignity and professional pride, general health problems, and prolonged stress (Table 5.10). Less common were low self-esteem and depression, which were also experienced by employees after incidents of aggression.

When asked about the most common ways of relieving stress after contact with an aggressive patient, respondents first indicated playing sports, listening to music, and meeting friends. Slightly less frequently, they indicated doing housework, watching television, calling a friend, and spending time with animals as ways to relieve tension from work. In addition to the methods mentioned, the respondents also indicated "other" activities, which included a wide variety of answers, for example: "I play video games," "I drink alcohol," "I forget about it right away," "I read a book," "I lock myself in my room," "I talk to my coworkers," "I joke about the situation," and "I talk about the situation at work," Respondents engage in various activities to

TABLE 5.11
Ways of Coping after an Incident of Aggressive Patient Behavior

Ways to Cope	N	%
Watching television	50	20.3
Listening to music	92	37.4
Being around animals	41	16.7
Playing sports	98	39.8
Meeting friends	75	30.5
Phone call to a friend	44	17.9
Household chores	54	22.0
Helping others	15	6.1
Other	43	17.5

Source: Based on survey results.

TABLE 5.12
Organizational Effects of Aggressive Patient Behavior

Organizational Consequences of Aggressive Behavior	N	%
Decrease in the quality of services provided	105	42.7
Decrease in work efficiency	134	54.5
Disruption of work	179	72.8
Risk of medical error	150	61.0
Staff absenteeism	48	19.5
Costs (renovation, purchase of new equipment)	85	34.6
Other	2	0.8

Source: Based on research results

relieve tension after incidents of violence at work, ranging from addictive behaviors to "dark humor." However, what can be noticed is the important role played by the opportunity to share a difficult experience with others, especially with people at work who understand the problem perfectly. In addition, joint analysis of a specific case allows for reflection in the search for the best courses of action and builds a sense of social support that is so important in such moments.

When asked about the organizational consequences of aggressive patient behavior in the ED, respondents most often cited disruption of work, higher risk of medical errors, decreased work efficiency, and reduced quality of service (Table 5.12). They were much less likely to point to costs associated with damage to medical equipment or other items in the ward or absenteeism.

Since the literature review identified problems with reporting incidents of patient violence, the survey asked respondents about the functioning of dedicated registers related to aggressive patient behavior. In the opinion of one-third respondents, Polish EDs have special registers for reporting aggressive patient behavior, both physical (34.9%) and psychological (36.9%). A slightly smaller number of respondents indicate the existence of registers for verbal aggression (28.2%). Similarly, about 30% of respondents admit that they have no knowledge in this area (Table 5.13).

TABLE 5.13
Presence of Records of Aggressive Behavior in Patients

	Records					
	Yes		No		I Don't Know	
Form of Behavior	N	%	N	%	N	%
Physical	86	34.9	82	33.3	84	34.3
Mental	91	36.9	100	40.6	88	35.7
Verbal	69	28.2	64	26.1	74	30.0

Source: Based on research results.

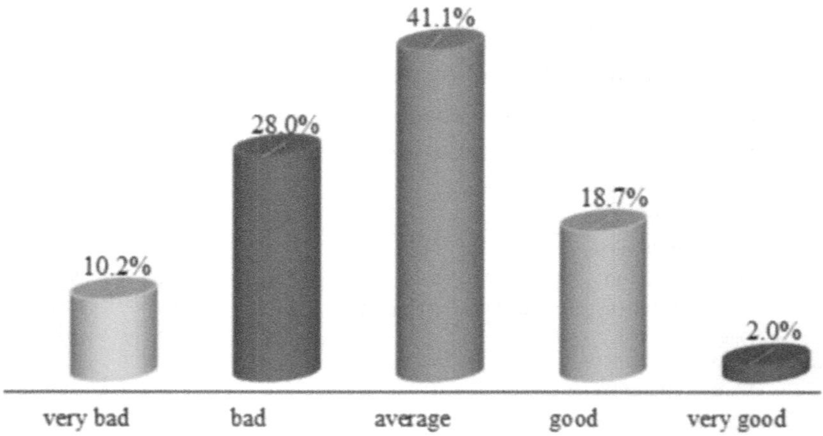

FIGURE 5.10 Preparedness of emergency departments to prevent aggressive patient behavior

More than half of doctors (60%) and paramedics (52.3%) confirmed that aggressive patient behavior was included as a risk factor in their "occupational risk assessment." The presence of aggressive patient behavior in the "occupational risk assessment" was also confirmed by almost half of the nursing staff surveyed (46.4%).

When it came to the preparedness of EDs to prevent aggressive patient behavior, most of the surveyed employees rated it as average (41.1%). Only 2% of respondents considered EDs to be very well prepared and 18.7% considered them to be well prepared to prevent aggressive patient behavior. On the other hand, as many as 10.2% noted that EDs are very poorly prepared and 28% noted that they are poorly prepared to deal with patient aggression (Figure 5.10).

According to the respondents, emergency departments do not have effective procedures for responding to aggressive patient behavior (37.0%), the causes of aggressive patient behavior are not analyzed in emergency departments (62.6%), and hospitals do not provide psychological assistance to victims of violence experienced from patients (92.3%). The surveyed staff believe that it is necessary to introduce content on dealing with aggressive patients into the training of medical personnel.

The collected data show that only 26.4% of employees benefited from the legal protection provided for public officials. When asked about possible ways to de-escalate violence in the ED, most respondents were in favor of establishing entities responsible for informing patients about the legal protection available to medical personnel (77.2%). Respondents agree that radical measures are needed to limit the escalation of aggressive behavior by patients (79.3%), pointing to the need for training in dealing with agitated patients, interpersonal communication, as well as applicable law and self-defense techniques (Figure 5.11).

In the survey, respondents also indicated the need for training for patients or campaigns to raise patient awareness about behavior in medical facilities, legal regulations, and especially their rights and other internal regulations in force in the ED.

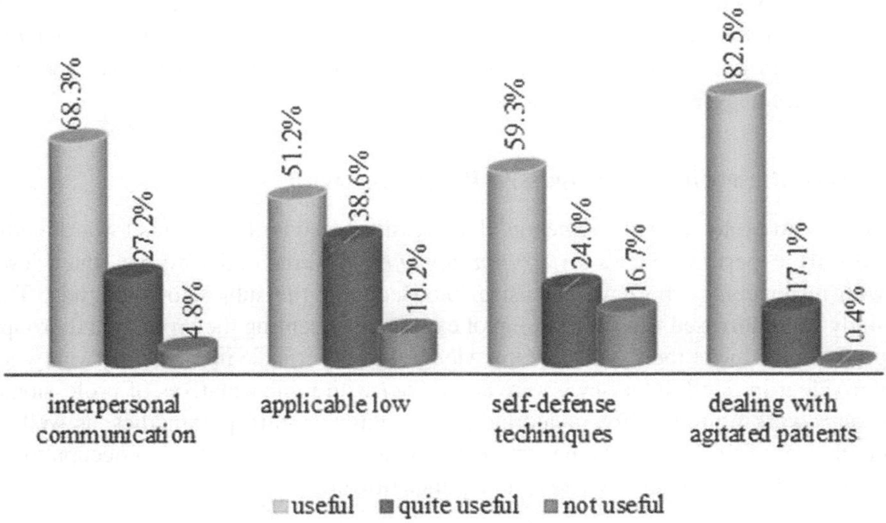

FIGURE 5.11 Training to increase employee safety

In summary, surveys conducted among emergency room medical staff indicated that the problem of aggressive patient behavior is a real threat in the workplace, occurring frequently and severely disrupting the work of the team. Medical staff are exposed to aggressive behavior in verbal, physical, and psychological forms. There are also cases of indirect aggressive behavior by patients using cell phones or the Internet. Most doctors, nurses, and paramedics experienced aggressive behavior by patients "once in a while" but some experienced such behavior "during every shift." The most common forms of aggressive behavior by patients were waving their arms, pushing, shoving, intimidation, threats, humiliation, and many forms of verbal aggression. Aggressive behavior was most often exhibited by patients under the influence of alcohol, less often by patients after using psychoactive substances (other than alcohol). The medical staff's response to aggressive behavior by patients was most often a request to change their behavior and a request for help from colleagues. Among the emotions experienced by employees after contact with a patient exhibiting aggressive behavior, anger, helplessness, and fear of the aggressor were most often reported. Unfortunately, a significant proportion of emergency room medical staff treat aggressive behavior by patients and their families as a permanent feature of their work. The lack of unanimity among respondents regarding the presence of records of aggressive behavior by patients, its various forms, and the identification of aggressive behavior by patients in occupational risk assessment charts is puzzling.

CONSEQUENCES OF AGGRESSIVE BEHAVIOR BY PATIENTS

The survey showed that aggressive behavior by patients toward medical staff in the ED poses a real threat. Employees experience such behavior from patients and their families very often, but these incidents are rarely reported. In order to understand this situation, it was decided to deepen the research and supplement the research material. To this end, two methods of qualitative data collection were used to explore the issue in greater depth: expert interviews and case studies of people who had experienced aggression from patients.

Experts' Perspective on Aggressive Patient Behavior

In order to explore the phenomenon of aggressive behavior by patients toward medical staff, expert interviews (*expert opinion survey*) were conducted, i.e., interviews with respondents who have extensive knowledge of the subject of the study. The study was addressed to a small group of experts representing the professional groups under study due to their specialist knowledge and experience in the field. Twenty-six individual in-depth interviews were *conducted* with representatives of professional groups working in the ED, including doctors, nursing staff, paramedics, as well as employees of the Quality Management Team and experts in the field of occupational health and safety, legal advisors, and a police officer.

Analysis of Expert Interviews

For the purposes of the study, an in-depth interview scenario was prepared, which included main questions along with supplementary questions. However, the questions were modified on an ongoing basis during the conversation with the respondent. The interviews lasted an average of about 30 minutes and were conducted in person by one of the authors. The situation of an individual conversation between the interviewer and the respondent allowed for greater openness and freedom of expression on this difficult topic. All respondents agreed to the use of the data obtained for scientific research, and the data obtained during the interview was recorded on an ongoing basis. The interviews were transcribed and their content was analyzed based on thematic analysis (in accordance with the procedure described by Braun and Clark) [2006, pp. 77–101]. Table 5.14 presents the characteristics of the expert sample studied.

Analysis of the information obtained during the interviews allowed us to conclude that the scale of aggressive behavior by patients toward medical staff is significant. Respondents stated that:

> Verbal aggression is so common that we treat it as a normal part of our work.
>
> *Doctor (R 1)*

The respondents' statements point to the sad fact that aggressive behavior by patients is an everyday occurrence in EDs. However, it is not possible to give a specific number of aggressive behaviors by patients toward ED medical staff due to the lack of an obligation to report such incidents in hospitals in Poland. As the respondents point out:

> ...patients who exhibit aggressive behavior are present on every shift, but it is not possible to provide the number of such incidents because there are no records of aggressive behavior by patients towards medical staff....
>
> *Nurse (R 8)*

TABLE 5.14
Characteristics of the Expert Sample

Symbol	Expert Position	Age (in Years)	Gender	Work Experience	Place of Work (Province)
R 1	Doctor	47	Male	21	Malopolonia
R 2	Doctor	50	Male	7	Masovian
R 3	Doctor	52	Female	26	Malopolonia
R 4	Doctor	41	Male	14	Masovian
R 5	Nurse	33	Female	18	Malopolonia
R 6	Nurse	58	Female	34	Malopolonia
R 7	Nurse	47	Female	23	Malopolonia
R 8	Nurse	29	Female	4	Silesia
R 9	Nurse	33	Female	7	Swiętokrzyskie
R 10	Nurse	31	Female	3	Pomeranian
R 11	Paramedic	31	Female	4	Silesia
R 12	Paramedic	29	Female	3	Malopolonia
R 13	Paramedic	36	Male	11	Malopolonia
R 14	Paramedic	28	Female	4	Malopolonia
R 15	Paramedic	41	Female	17	Subcarpathian
R 16	Paramedic	46	Male	19	Pomeranian
R 17	Chair of the Quality Management Team	50	Male	24	Malopolonia
R 18	Member of the Quality Management Team	34	Female	5	Malopolonia
R 19	Health and Safety Specialist	69	Male	36	Malopolonian
R 20	Health and Safety Specialist	55	Male	27	Subcarpathian
R 21	Health and Safety Specialist	37	Male	9	Lower Silesia
R 22	Health and Safety Specialist	39	Male	13	Malopolonia
R 23	Health and Safety Specialist	33	Female	7	Masovian
R 24	Police Officer	43	Male	16	Silesia
R 25	Legal Advisor	52	Male	23	Malopolonia
R 26	Legal Advisor	29	Female	4	Malopolonia

Source: Study based on research results.

And further,

> …in the emergency room, there are cases of aggressive behavior by patients towards medical staff on every shift, but it is difficult to provide the number of incidents due to the lack of reporting of such incidents.…
>
> *Paramedic (R 14)*

The aforementioned lack of mechanisms for reporting incidents of aggressive behavior by patients in the ED means that the data is fragmentary. Information on aggressive behavior by patients is recorded in electronic medical records, which are only accessible to immediate supervisors:

> ...Information about aggressive patient behavior is only recorded in electronic nursing reports, which are accessible only to the ED physician and the ward nurse....
>
> *Nurse (R 9)*

Respondents indicated that the lack of reporting and recording of such incidents in the documentation is due to the lack of formal regulations requiring such an obligation and the attitudes of supervisors and other employees at the facility. Even incidents that meet all the conditions necessary to be considered an accident at work are not reported, as evidenced by the following statement from one of the experts.

> ...when I took "protective" measures, i.e., connecting the IV, I was bitten on my left forearm. In my opinion, the injury was not serious, because the skin was not broken and the only consequence of the incident was a large bruise. I did not report the incident anywhere or make a note of it, because I am not obliged to do so and no one encourages me to report such incidents....
>
> *Paramedic (R 13)*

As can be seen, the attitudes of other employees also do not encourage the reporting of aggressive patient behavior, and in addition, there is a disturbing habit associated with the presence of aggressive patient behavior, which is becoming a norm in the workplace. This may show the powerlessness of employees and their acceptance of this state of affairs, as well as their tacit consent to aggressive patient behavior.

In addition to the relevant legal acts, there is also a lack of internal regulations that would require employees to report incidents of aggressive patient behavior in the emergency room.

> ...there are no legal provisions that would oblige employees to report incidents involving aggressive patient behavior. Such incidents are most often reported by staff to their immediate superiors, who in reality have no power to do anything about them.
>
> *Doctor (R 1)*

The feeling of powerlessness of immediate superiors in the face of such incidents is also disturbing. Respondents feel helpless in the face of aggressive behavior by patients in the ED, which further discourages them from reporting such incidents in medical records and to public authorities.

> ...ER employees refrain from reporting incidents of aggressive patient behavior to public authorities because of their powerlessness....
>
> *Doctor (R 2)*

Information related to aggressive behavior of patients toward medical staff in the ED is usually included in nursing reports, which are kept in accordance with applicable laws on electronic medical records.

> …incidents of aggressive behavior are recorded in the nursing report and as adverse events in the so-called adverse event cards and are forwarded to the Quality Management Team….
>
> *Paramedic (R 15)*

In view of the paramedic's statement, the question arises as to whether incidents of aggressive behavior by patients toward medical staff can be classified as adverse events. Unfortunately, according to the definition and classification, **an adverse event** is damage to the patient's health caused during treatment that is not related to the course of the disease or the patient's state of health. *Chair of the Quality Management Team at the hospital (R 17).* Therefore, it cannot be classified *as an adverse event.* "Adverse events" include events that lead to harm or may cause such harm only to patients – not to employees. Incidents involving aggressive patient behavior often meet the conditions necessary to classify events as occupational accidents and should be reported to the health and safety department.

According to respondents, ED medical staff are increasingly experiencing physical forms of aggressive behavior from patients, while verbal forms of aggressive behavior occur on every shift and are treated as an integral part of working in the ED.

> …aggressive behavior by patients in the ED occurs very often, on every shift. Most often, it is verbal behavior, which we treat as part of our job. Increasingly, there are aggressive behaviors in the form of physical attacks. ..
>
> *Paramedic (R 11)*

Contradictory information was obtained regarding aggressive behavior by patients in the EDduring the SARS-CoV-2 epidemic. On the one hand, the scale of aggressive behavior by patients in the ED decreased due to the restrictions introduced, which prevented patients' families and other accompanying persons from staying in the ward.

> …we are seeing fewer incidents of aggressive behavior by patients because, due to the risk of coronavirus infection, a ban or significant restrictions have been introduced on accompanying persons, including family members, staying in the ED….
>
> *Nurse (R 7)*

On the other hand, drawing the attention of patients who did not comply with the regulations on wearing a mask covering their mouth and nose resulted in an increase in incidents of aggressive behavior on their part, mainly in verbal form. A significant proportion of patients came without a mask covering their mouth and nose. Requests to wear a mask often resulted in the destruction of ward equipment, especially medical equipment.

> We are afraid for our own health, so we try to enforce the wearing of masks by patients. However, the effect is often the opposite; instead of putting on a mask, the patient reacts aggressively: he insults, threatens, threatens to beat us up, and makes vulgar and arrogant remarks. I remember a patient who reacted aggressively to a request to wear a mask, destroying several chairs and drip stand....
>
> *Paramedic (R 14)*

In addition to physical forms of aggression, patients also exhibited other forms of aggressive behavior.

> Patients behave aggressively when they are told they are not wearing a mask, they swear, insult, intimidate, threaten....
>
> *Nurse (R 8)*

Aggressive behavior by patients in the emergency room resulting from being told they were not wearing a protective mask was not an isolated incident, and the lack of sanctions did not encourage compliance with regulations on wearing protective masks.

> ...there is an obligation to wear masks, but patients ignore this obligation. When they are told about it, they behave aggressively. The case ended up in court, but was dismissed due to the low harmfulness of the act....
>
> *Doctor (R 1)*

There were legal grounds for imposing penalties on people who refused to comply with the regulations requiring the wearing of protective masks and on people who violated the personal rights of others, but in many cases the penalties imposed were grossly disproportionate to the acts committed.

> The patient was asked to put on a protective mask in accordance with the applicable restrictions. He refused and started kicking the door, broke into the doctor's office where I was staying, and punched me in the face. As a result, the aggressor was punished with police supervision and a restraining order prohibiting him from approaching or contacting me (the victim)....
>
> *Paramedic (R 11)*

When asked about **the reasons for aggressive behavior in patients**, medical staff pointed to the behavior of patients brought to the emergency room by the emergency medical services. Patients brought to the emergency room were more likely to exhibit aggressive behavior, which was independent of their will and could occur for many reasons, including being in a confined space.

> ...I believe that their aggressive behavior is exacerbated by the confined space of the ambulance....
>
> *Paramedic (R 15)*

The emergency room also had so-called uncooperative patients who got out of bed, walked around the rooms, and interfered with work. The second group consisted of patients who exhibited aggressive behavior. Among them were people under the influence of alcohol, psychoactive substances, mainly drugs, and people with mental disorders. The most common cause of aggressive behavior in patients was alcohol intoxication, but there were situations in which the patient's state of sobriety was misjudged and, as a result, the causes of aggressive behavior were misidentified.

> …it happens that the cause of a patient's aggressive behavior is misjudged by questioning their sobriety. I remember a situation in which, after conducting a laboratory test for the presence of alcohol in the patient's blood, the result was negative, which ruled out alcohol consumption as the cause of aggressive behavior. Tests were also conducted for the presence of narcotics and drugs in the blood. No such substances were found in the patient's blood. The doctor on duty decided to perform a CT (computed tomography) scan of the patient's head, which revealed a craniocerebral injury, specifically a hematoma. This was most likely the cause of the patient's aggressive behavior….
>
> *Nurse (R 6)*

The emergency room medical staff did not always have the opportunity to verify their knowledge of the causes of the patient's aggressive behavior. Such situations occurred when a patient was admitted at the end of the day shift and then transferred to the next shift – the night shift.

> …my colleague, a nurse who worked during the day, is unable to verify her knowledge of the causes of the patient's aggressive behavior. Information about the patient's health and the causes of his aggressive behavior, confirmed by diagnostics, can be found in the medical records ( ). There is no reason to review the patient's medical records on subsequent working days resulting from the shift schedule, so there is no certainty that the assessment of the cause of the patient's aggressive behavior was correct….
>
> *Nurse (R 7)*

Another cause of aggressive behavior in patients in the emergency room, which was pointed out by experts, was nicotine craving. Nicotine is a substance that stimulates receptors in the brain. Its effect is an increased secretion of dopamine, which is responsible for feelings of pleasure, joy, and happiness. Regular delivery of nicotine to the body causes the brain to become accustomed to the presence of nicotine and, as a result, to tolerate it. Waiting for hours for advice in the ED and abstaining from smoking causes nicotine cravings, which result in aggressive behavior.

> …nicotine cravings are a common cause of aggressive behavior in patients. A total ban on smoking in hospital buildings and long waiting times to see a doctor in the emergency room cause patients who are heavy smokers to experience nicotine withdrawal, with symptoms such as fatigue, low mood, concentration problems, irritability, and aggressive behavior….
>
> *Health and Safety Specialist (R 20)*

Patients under the influence of alcohol and patients with mental disorders who are also under the influence of alcohol are a cause for concern. Depending on their blood alcohol concentration, patients are placed in specialized medical facilities that treat mental disorders and poisoning with strong intoxicants. The police are responsible for transporting patients to sobering-up centers, while medical transport is used to transport patients to other healthcare facilities.

> …most often, patients under the influence of alcohol and patients with mental disorders under the influence of alcohol behave aggressively. Patients under the influence of alcohol are placed in a sobering-up center, while patients with mental disorders are placed in a psychiatric hospital. However, the placement of a patient in the above-mentioned facilities depends on the blood alcohol concentration. Patients with a blood alcohol concentration not exceeding 3.00‰ are admitted to a sobering-up center, while patients with a blood alcohol concentration not exceeding 1.00‰ are admitted to a psychiatric hospital. The police only transport patients to sobering-up stations in cases where there is no immediate threat to the patient's life or health and the blood alcohol concentration does not exceed 3.00‰. However, we do so reluctantly.
>
> *Police Officer (R 24)*

The interviews also showed that doctors, nursing staff, and paramedics at the ED did not take advantage of the protection afforded to public officials. They did not report incidents of aggressive behavior by patients to the police due to formalized procedures requiring time and significant involvement, as well as the associated costs. The main factor justifying this approach was the discomfort associated with leaving the workplace in order to summon the victim to court and the excessive workload placed on colleagues remaining at the workplace.

> …I do not take advantage of my status as a public official and do not report incidents of aggressive behavior by patients to the state authorities because the procedures require a great deal of involvement, especially in court proceedings. Participation in court proceedings is important for the resolution of the case, but it involves additional costs, e.g., travel, parking, etc. Due to the shift system in the emergency room, attendance at a court hearing often requires sacrificing free time intended for rest. In a situation where, according to the work schedule, the summons concerns a day on which I am supposed to work, my employer exempts me from the obligation to work for the time I spend in court, while maintaining my right to remuneration, but does not cover the related costs. In addition, leaving my workplace for the duration of the commute and court appearance causes me discomfort, because my absence, on the one hand, reduces the number of staff, and on the other hand, causes more work for the employees remaining at the workplace, thus increasing the waiting time for patients to be admitted and undergo medical procedures….
>
> *Paramedic (R 16)*

The law provides protection for doctors, nursing staff, and paramedics in the emergency room.[1] However, employees do not take advantage of this protection due to unfriendly and bureaucratic procedures, including the need to prepare documents containing their personal data, which may be made available to the perpetrator of the incident in criminal proceedings.

> ...staff are reluctant to take advantage of the legal protection available to public officials and are reluctant to report aggressive behavior by patients. The current reporting form used by police officers contains not only the personal data of the perpetrator, but also the personal data of the victim, to which the perpetrator may have access. In order to protect the victim from violence, unlawful threats, or other behavior intended to intimidate or retaliate, it is possible to keep personal data, including place of residence and place of work, confidential. However, this requires submitting a request to the public authorities to keep the victim's personal data confidential, which entails additional costs – travel to the institution and spending one's free time, etc.
>
> *Health and Safety Specialist (R 23)*

Due to the aggressive behavior of patients, the police are often called to the emergency room. Unfortunately, there have been cases of refusal to take action and unfounded accusations of unjustified intervention. Refusal to intervene and unfounded accusations cause indifference and a lack of response from medical staff to aggressive behavior by patients.

> ...the patient, who was under the influence of alcohol and staying in the emergency room, behaved aggressively. The police were called, but they refused to take action in connection with the patient's aggressive behavior. The medical staff of the hospital's emergency department were accused of making an unjustified call, and the hospital was obliged to cover the costs of the allegedly unjustified call. In order to punish the perpetrator for his aggressive behavior in the emergency room, the police officers recommended filing a motion with the court and initiating civil proceedings. The hospital filed a complaint with the Police Headquarters regarding the behavior of the officers. Unfortunately, the response the hospital received from the Police Headquarters was that no violations had been found, which is why we very rarely report incidents involving aggressive behavior by patients due to unfounded accusations....
>
> *Doctor (R 1)*

Unfortunately, the accusations of unjustified calls were confirmed by persons providing legal assistance to medical facilities.

> ...I confirm the veracity of the described situation of an allegedly unjustified call to the police regarding an aggressively behaving patient in the emergency room....
>
> *Legal Advisor (R 26)*

Respondents only turned to the police for help in cases of physically aggressive patient behavior in order to avoid financial liability. In order to limit the escalation of other forms of aggressive patient behavior, employees tried to deal with aggressive patient behavior on their own.

> ...I ask for help when property is damaged or destroyed. I do this to avoid financial liability for destroyed or damaged equipment....
>
> *Nurse (R 8)*

The respondents' statements indicate a low awareness of the harmfulness of the problem of aggressive patient behavior and **a lack or low level of cooperation between key social partners** (stakeholders) in relation to unethical patient behavior.

The lack of legal acts and internal regulations containing provisions requiring the punishment of perpetrators of incidents involving aggressive behavior causes **indifference and a lack of response from staff** to the escalation of aggressive behavior by patients.

> ...I do not report incidents involving aggressive patient behavior because there are no legal provisions or internal procedures that would require such incidents to be reported to anyone....
>
> *Paramedic (R 12)*

Aggressive patient behavior is not a reason to report the incident to the health and safety department. The reporting obligation applies to specific incidents that meet the conditions for an incident to be considered an accident at work or a near miss.

> ...I do not report incidents of aggressive patient behavior to the health and safety department because there is no such obligation....
>
> *Paramedic (R 16)*

Interviews with health and safety experts showed that injuries resulting from aggressive patient behavior are sometimes reported to the health and safety department. Although they are one of the necessary conditions for an incident to be considered an accident at work and for benefits to be obtained, they are reported only sporadically.

> ...an employee suffered a broken nose as a result of aggressive patient behavior. It happened that the victim reported the incident, which was recorded as an accident at work, but I don't remember in which ward of the hospital the incident took place. I carried out all the procedures related to the accident at work, drew up an accident report and a statistical accident card, and forwarded the documents to the Social Insurance Institution in order for the victim to obtain compensation for the accident at work. I am not aware of the further course of the case....
>
> *Health and Safety Specialist (R 21)*

There are no national regulations in Poland that address the phenomenon of aggressive patient behavior, so there is no obligation to include it in occupational risk assessment cards. In addition, the lack of statistical data on the frequency of aggressive behavior by patients toward medical staff makes it impossible to assess the level of risk associated with this phenomenon. This, in turn, translates into a lack of measures to limit the escalation of aggressive behavior by patients.

> ...aggressive behavior on the part of patients is a professional risk factor for all hospital employees who have contact with patients, regardless of their place of work. The occupational risk assessment does not take this risk into account because aggression

> is not listed as a risk factor in any national regulations. There is a lack of statistical data that could be used to carry out an occupational risk assessment, which is why no measures are taken to prevent or limit the escalation of aggressive behavior by patients, especially in the emergency room....
>
> *Occupational Health and Safety Specialist (R 19)*

The occupational risk assessment of professional groups that have direct contact with patients should take into account the risk of exposure to aggressive behavior on the part of patients, the forms of this behavior, and the level of risk at each workplace.

> ...The occupational risk assessment includes the risk of aggressive behavior on the part of the patient in general, without taking into account specific forms of aggression or the level of this risk. It is prepared for professional groups and not for workstations. Creating separate occupational risk assessments for workstations in individual hospital wards would result in a significant expansion of health and safety documentation....
>
> *Health and Safety Specialist (R 22)*

The group of experts, also included legal advisors from one of the hospitals, that provide comprehensive assistance in matters related to the hospital's activities. One of the areas of activity of the legal advisors is proceedings to punish the perpetrator (patient) in connection with his or her aggressive behavior during their stay in the emergency room. Information was obtained that proceedings had been initiated in relation to all reported incidents of aggressive behavior by patients in the emergency room, but it was not possible to obtain detailed information due to the obligation of professional secrecy.

> ...several incidents of aggressive behavior by patients in the ER are reported annually in the hospital, and these are only incidents that took the form of physical attacks. Due to professional confidentiality, I cannot provide any further information or disclose any documentation. The law requires us, as legal advisors, not only to maintain professional confidentiality, but also to protect against the disclosure or misuse of information related to the performance of our professional duties....
>
> *Legal Advisor (R 25)*

According to the law, employers should ensure the safety of their employees at work and take measures to minimize the risk of health hazards, including those resulting from aggressive behavior by patients. Unfortunately, research has shown that employers sometimes fail to take such measures in the case of aggressive patient behavior due to a lack of professional tools that would allow for effective action to reduce the risk associated with aggressive patient behavior, especially in emergency departments.[2]

> ...the employer is obliged to take measures to ensure the safety of employees in the workplace, but in the case of aggressive behavior by patients, no action is taken....
>
> *Doctor (R 2)*

Despite their obligations to ensure safe and hygienic working conditions, employers are unable to prevent medical staff from being exposed to aggressive behavior on the part of patients. However, they do take care of property for which the employee is financially responsible.

> A reaction only occurs when property is damaged. In the event of damage to property, every employee reports such an incident to their supervisor so as not to bear the costs associated with it. When physical acts of aggressive behavior occur, resulting in injury, I have to take care of my own health. I have no difficulty with this, because I can get medical help during work….
>
> *Paramedic (R 14)*

Emergency room staff receive the necessary medical assistance in the event of aggressive behavior by a patient directly at their workplace and do not report incidents related to aggressive behavior by patients due to a lack of faith in the effectiveness of intervention.

> …if there is damage to health, depending on its scale, I either seek help from the doctor on duty in the emergency room or remain indifferent to the situation, because it won't change anything anyway….
>
> *Paramedic (R 16)*

Depending on the form of aggressive behavior of the patient and the effect of this behavior, the perpetrator is liable under specific provisions of the criminal code. In the opinion of the respondents, the current legal provisions are insufficient, which is why new legal regulations are necessary to protect medical personnel from aggressive behavior on the part of patients, especially in the ED. The penalties imposed on patients who exhibit aggressive behavior toward medical personnel in the ED are unsatisfactory.

> …for aggressive behavior towards medical staff, a patient may be punished under specific provisions of the Criminal Code, but in my opinion, the penalties applied are insufficient….
>
> *Doctor (R 1)*

There is an urgent need for legislative work[3] aimed at regulating matters related to limiting the escalation of aggressive behavior by patients and updating the provisions on the protection of public officials.

> …in my opinion, in order to limit the escalation of aggressive behavior by patients towards medical personnel, new legal regulations are necessary, especially in the area of protection afforded to public officials….
>
> *Doctor (R 2)*

During expert interviews, a case was identified of a patient physically attacking a paramedic providing medical assistance in the emergency room. The paramedic, defending himself, reacted in a similar manner by striking the patient, which was recorded by surveillance cameras. As a result of the situation, the paramedic suffered serious consequences, which he did not want to disclose to the interviewer. He added, however, that the patient's behavior went unpunished.

The expert interviews helped to understand the essence of the problem, especially the lack of reporting of such incidents by medical staff in the emergency room. Unfortunately, they highlighted the lack or ineffectiveness of measures taken to ensure the safety and health of medical staff and the imperfection of the legal system in counteracting aggressive behavior by patients in the emergency room. The phenomenon of aggressive behavior on the part of patients is not mentioned in the law, and the inclusion of this risk in the occupational risk assessment charts of medical personnel is voluntary. The research showed a lack of statistical data on incidents related to aggressive behavior of patients toward medical personnel, which results in a lack of risk assessment related to such behavior. There is also a lack of records of aggressive behavior by patients in EDs and tools that would limit the escalation of aggressive behavior by patients, and the penalties imposed on aggressors are unsatisfactory. Despite the fact that the law provides protection for doctors, nursing staff, and paramedics in the emergency room, similar to that provided for public officials, employees are reluctant to use this protection, not believing in the effectiveness of such measures. One may wonder why employees do not take action to change this difficult situation and break the silence surrounding this difficult and harmful behavior of patients. Perhaps such behavior is a manifestation of the so-called **organizational silence** that occurs in healthcare facilities?

Organizational Silence in Medical Facilities

Organizational silence was noticed and introduced into science in 2000 by Elizabeth W. Morrison and Frances J. Milliken and is understood as a group phenomenon that occurs in situations where the majority of members of an organization do not speak out on matters concerning the organization [Adamska 2016, p. 77]. The phenomenon consists of not speaking up in situations that are unfair, illegal, or immoral that occur within the organization. Authentic communication does take place in such organizations, but it happens "behind closed doors" so that no one finds out about it [Cewińska, 2016]. **Organizational silence** is a resignation from the possibility of influencing a situation that is considered undesirable and manifests itself in a lack of faith in the possibility of change and corrective action [Adamska 2016, p. 75; Kudryńska, Krzywdziński 2021, p. 123]. Unfortunately, in the long run, it can lead to an accumulation of problems and conflicts, lower productivity, lack of commitment, or employees leaving their jobs.

Management literature identifies three types of organizational silence:

- Acquiescent silence,
- Defensive silence,
- Prosocial silence.

Conciliatory silence occurs when employees remain silent because they do not believe their arguments will be effective. Defensive silence occurs when an employee remains silent in order to protect themselves. Prosocial silence, on the other hand, refers to situations where silence results from cooperation for the good of the organization [Dyne, Ang, Botero 2003, p. 1359; Kudryńska, Krzywdziński 2021, p. 124]. The following factors play a key role in maintaining silence: anticipation of negative consequences of speaking up, an unfavorable organizational climate, belief that speaking up will not change anything, individual characteristics of employees, and lack of support and trust from superiors [Kudryńska, Krzywdziński 2021, pp. 123–124].

The phenomenon of organizational silence in healthcare facilities is described in the context of medical errors. Medical errors are one of the most serious problems due to preventable deaths and serious adverse events, which are highly emotional due to their tragic nature [Henriksen, Dayton 2006, p. 1541]. In this case, the phenomenon of organizational silence is defensive and pro-social in nature. Employees refrain from speaking out in order to protect themselves, shift suspicion onto someone else, or strengthen the image and avoid unpleasant financial consequences for the organization.

In relation to aggressive behavior by patients, the phenomenon of organizational silence is conciliatory in nature. Employees do not talk about aggressive behavior on the part of patients, suppress their concerns about work safety, and distance themselves from taking actions that could ensure their safety at work and high-quality care for patients. They are convinced that talking about the problem will not change anything and may only result in negative consequences such as job loss, being passed over for promotions and rewards [Kudryńska, Krzywdziński 2021, p. 124]. Employees also remain silent due to a lack of confidence or fears about the consequences of their involvement. The failure to disclose relevant information and opinions results in a lack of action to limit the escalation of aggressive behavior by patients. Employees suppress their agitation and distance themselves from discussing issues related to aggressive patient behavior, thereby failing to take actions that could ensure their safety at work. Incidents involving aggressive patient behavior are both recognized and concealed. If incidents involving aggressive patient behavior are not reported, knowledge about them remains inaccessible and safety issues related to aggressive patient behavior will continue to be overlooked in the management of medical facilities [Henriksen, Dayton 2006, pp. 1539–1540].

Both within and outside the healthcare sector, the phenomenon of organizational silence is dangerous, as it causes a number of negative effects for both employees and organizations, including: reduced engagement, inhibition of development and innovation, suffering and violence, sexual harassment, and medical errors.

Therefore, it seems crucial not only to diagnose and assess the scale of the problem of aggressive behavior by patients toward medical staff but above all to raise awareness of this problem among various social partners. Especially among decision-makers involved in the management of such organizations. To this end, cases of emergency room medical staff who have experienced particularly severe negative consequences of aggressive patient behavior in their work were analyzed. This behavior had a significant impact on their future lives and professional careers.

Analysis of Particularly Severe Cases of Aggressive Behavior by Patients toward Medical Staff

Expert interviews allowed us to learn about cases and reach people working in medical facilities who were particularly affected by aggressive behavior on the part of patients. This influenced the further exploration of these cases and the conducting of interviews with people affected by aggressive behavior on the part of patients. In order to understand and analyze such incidents, the *case study research* method was used. *A case study* involves the examination of phenomena and processes in their real environment and a comprehensive description of these phenomena [Czakon 2022, p. 208].

At this stage of the research, 13 cases of aggressive patient behavior were analyzed, and the collected qualitative data allowed for the classification of the presented cases in terms of their consequences for:

- Further professional career.
- Impact on the health of medical staff exposed to aggressive patient behavior.
- A strong sense of social injustice.

Medical personnel are increasingly exposed to aggressive behavior, which negatively affects their future professional careers. Experiencing aggressive behavior from patients generates costs related not only to changing jobs and leaving the profession but also to staff absenteeism due to illness, which is particularly worrying in the context of the unfavorable age structure and staff shortages in this sector.

Description of Cases That Have a Negative Impact on the Careers of Medical Staff

Case Description 1 (C1). Nurse Harassed through the Disclosure of Personal Data (R 8)

Medical facilities have an identification requirement, which consists of the obligation to wear a badge in a visible place containing the employee's first and last name and position.[4] A patient staying in the emergency room during medical treatment used his own cell phone to take a photo of a nurse and read her personal data from her ID badge: her first and last name and her position. After leaving the emergency room, he regularly visited the ward in the following days and looked for the woman, giving her first and last name and showing her a photo taken with his own cell phone. In direct contact, he blackmailed and threatened the woman and her family. The victim filed a private indictment against the patient in court. However, in order to avoid further contact with the persistent patient, which intensified her fear and anxiety, she decided to resign from her job at this medical facility. The further course of the proceedings is unknown due to the lack of contact with the victim.

Case Description 2 (C2). False Accusation against a Doctor (R 3)

A physician – a long-standing specialist – performed medical procedures on a minor patient with due diligence. However, the minor's legal guardian posted a false accusation on an Internet forum, providing the personal details of the physician who had provided the care. The incident was reported to the doctor in charge of the ED, who carefully checked the monitoring and patient records and found no irregularities. As part of the intervention, both the hospital management and the doctor in charge of the department submitted a request to the IT company to remove the doctor's personal data and false information. The hospital spokesperson and the legal team also looked into the matter. So far, only one response has been received – a refusal. The post has not been removed, and the doctor, fearing further false comments, terminated his employment contract and changed jobs.

Case Description 3 (C3). Doctor Slandered Online (R 4)

On a website for searching and evaluating medical specialists in Poland, a patient described her visit to the emergency room, presenting the doctor by name as a person who "took it out on her." She wrote hurtful and offensive comments and threats about the doctor. Within a short time, patients routinely gave up waiting for a doctor's appointment as soon as they learned that the doctor was on duty in the emergency room. Patients also wrote complaints to the management, which resulted in the termination of the contract between the hospital and the doctor for the provision of medical services in the emergency room.

Conclusions from the Analyzed Cases

One of the factors influencing the course of a professional career are posts on Internet portals evaluating the work of medical personnel. Slander and **aggression on the Internet** violate the good name and dignity of medical personnel, especially doctors. The posts often contain offensive language and false information, which result in people leaving their jobs and cause a significant deterioration in the medical services market (C1, C2, and C3). In most cases, slandered medical staff are unable to confront the hater. The lack of response from IT companies to such posts is highly disturbing, as it encourages patients and their families to express their subjective feelings and grievances toward medical staff.

Under current law, it is very difficult to hold online aggressors accountable, especially in the case of people threatening doctors. An additional difficulty is that online aggressors are anonymous and difficult to identify. Every incident of aggressive behavior expressed online should be reported to the police. Only then is there a chance that the posts on social media will be removed.

Description of Cases of Aggressive Behavior by Patients That Has a Negative Impact on the Health of Medical Staff

Aggressive behavior by patients in the emergency room is a phenomenon that negatively affects not only the course of their professional careers but also the health of employees and their personal lives. Health is a paramount value that must be protected and sought after. Establishing proper relationships with patients under the influence of alcohol is difficult due to the effect of alcohol on the central nervous system, during which the patient experiences inhibition of negative feelings and increased aggression. In situations of direct contact with a patient under the influence of alcohol, emergency room staff, whose role is to save the patient's life and health, often suffer damage to their own health.

Case Description 4 (C4). Paramedic Kicked in the Face (R 12)

According to the account of the injured paramedic, while performing medical procedures, an aggressive patient under the influence of alcohol kicked him in the face. As a result of this incident, the paramedic suffered a broken nose and damage to the lower alar cartilage. Medical assistance from an otolaryngology specialist was necessary. According to the paramedic, the incident resulted in permanent bodily injury in the form of deformation of the bridge of the nose and necrosis of the cartilage. In this situation, the police were called to deal with the aggressive patient, and they filed a motion to punish the perpetrator for assaulting a public official and causing bodily injury. After several court hearings, in which the paramedic had to participate, the perpetrator was punished, but only with a suspended sentence, which did not satisfy the injured paramedic.

Case Description 5 (C5). Paramedic Injured with Scissors (R 11)

A paramedic was injured while trying to restrain an aggressive patient. The victim did not notice when the patient took a pair of scissors and stabbed him in the back. He did not experience the incident because adrenaline, known as the stress hormone, significantly increased his pain threshold. The incident resulted in an injury that required medical attention from a specialist surgeon. According to the victim, "the result of this incident is a scar on my back and trauma that will remain with me for the rest of my life."

Case Description 6 (C6). Nurse Struck with a Belt (R 8)

A nurse who had been working in the emergency room for over 20 years was struck with a restraint belt during an intervention. The belt is made of natural leather and is used to restrain the arms and legs. At one end, it has a leather "connector" with a metal buckle, which is used to attach it to the bed frame, and the other end of the belt is a clamp for the limbs. The patient struck the nurse with the metal buckle, causing her to suffer facial injuries. The victim received medical assistance in the emergency room and did not report the incident to anyone because, as she claims, she is not obliged to do so.

Case Description 7 (C7). Paramedic with Suspected HIV Infection (R 13)

An aggressive patient under the influence of alcohol had problems to maintain his balance but tried to get out of bed. The paramedic attempted to keep the patient in bed. As a result of this interaction, he noticed signs of blood on his protective gloves. According to the paramedic's account, the blood was of unknown origin. He suspected that either he or the patient could have been injured during contact with the patient. In order to treat the injury properly, a post-exposure procedure was initiated, the purpose of which is to perform serological testing for HBV, HCV, and HIV infection in the exposed person and the person responsible for the exposure. As a result of testing the person responsible for the exposure (the patient), a positive result for HIV antibodies was obtained. In accordance with the applicable post-exposure procedure, within 24 hours of the incident, the injured person reported to the Post-Exposure Clinic, with which the hospital had concluded a contract for post-exposure treatment services. At the clinic, it turned out that the patient had been HIV-positive for many years but had not yet undergone treatment. By decision of a doctor specializing in infectious diseases, the paramedic did not receive antiretroviral drugs. However, he remained fearful and anxious about infection, which accompanied him throughout the following year, until he underwent further tests in accordance with the post-exposure procedure guidelines.

Case Description 8 (C8). Paramedic Struck in the Face (R 15)

The patient, who was lying down, calm, and not exhibiting aggressive behavior, punched a paramedic who was performing medical procedures in the face. As a result of this incident, the victim suffered numerous superficial injuries with massive swelling of the nose and damage to the blood vessel walls, resulting in nosebleeds. The paramedic received medical assistance in the emergency room, and the incident was not reported to anyone or recorded in any documentation due to the large number of procedures involved in punishing the perpetrator.

Case Description 9 (C9). Doctor Attacked with a Fire Extinguisher (R 2)

A patient under the influence of psychoactive substances unexpectedly got out of bed and activated a powder fire extinguisher installed on the wall of the emergency room corridor. Activating the extinguisher caused the fire extinguishing powder (a phosphorus salt compound) to be ejected under high pressure. The sprayed powder reduced visibility and severely irritated the doctor's eyes. Additional tests showed that the powder from the extinguisher had settled in his respiratory tract, making it very difficult for him to breathe. Attempts at self-treatment proved ineffective. The injured doctor required medical assistance (intervention by an ophthalmologist and pulmonologist), including a series of specialist tests. As a result of his injuries, he was unable to work and was on sick leave for several weeks.

Conclusions from the Analyzed Cases

Acts of physical aggression occur in the ED, which significantly affect the ability to perform professional duties and require specialist medical assistance (C4–C9). These incidents not only cause physical damage but also leave a mark on the psyche. Long-term employment and professional experience do not protect employees from aggressive behavior on the part of patients. There is still a lack of legislative solutions aimed at protecting medical staff, their lives, and health from the effects of aggressive patient behavior in some countries. Unfortunately, the inability to adequately punish the perpetrator increases the sense of helplessness among the affected employees. Employees' concerns about lengthy procedures for reporting incidents and their further processing, combined with a lack of confidence in their effectiveness and overly lenient penalties, are the main reasons for the lack of reports of patient aggression in emergency departments.

The health effects of medical staff coming into contact with aggressive patients are not always visible. The effects of tissue damage and contact with the biological material of an infected patient are delayed. Exposure to HBV, HCV, and HIV viruses is particularly dangerous. Symptoms of infection appear only after several weeks or more, and tests to detect infection are positive approximately 3 months after exposure. Therefore, post-exposure management is a very stressful situation due to the duration of the "serological window" (C7).[5]. The stress and trauma associated with such events remain for life and accompany employees throughout their careers, strongly influencing their work and commitment.

Currently, patients use many psychoactive substances, some of which are used in medicine as strong painkillers. They affect the central nervous system, causing mood swings from euphoria to depression, apathy, irritability, and even tantrums. Their effects also cause changes in perception, consciousness, and aggressive behavior, which is why ED staff should be particularly cautious when dealing with such patients.

Description of Cases of Aggressive Patient Behavior with Negative Social Consequences

Some manifestations of aggressive behavior in patients can result not only in the loss of health of medical staff but also have far-reaching consequences for the lives and functioning of the families of these employees, even though they are not directly involved in the provision of medical services. Therefore, when pointing out the negative social consequences of aggressive behavior in patients, the consequences for the families of employees and third parties are primarily indicated.

Case Description 10 (C10). Nurse Kicked in the Stomach (R 10)

A 21-year-old female patient was brought to the emergency room by the emergency medical services, accompanied by the police. She had escaped from a psychiatric hospital, where she was being treated for auditory hallucinations. The patient was agitated, exhibited aggressive behavior, and expressed a desire to "end her life" or kill other people. When the nurse was taking medication out of the cabinet, she had her back turned to the patient's bed. The patient suddenly got out of bed, approached the nurse standing with her back turned, and as soon as she turned around, she was pushed against the cabinet door and fell to the floor. A moment later, the patient kicked the nurse in the stomach. The nurse was in the early stages of pregnancy. The injured nurse claims that as a result of the patient's aggressive behavior, she lost her pregnancy. The case was taken up by the police, who were at the scene. The nurse was issued a sick leave certificate and the patient was placed in a psychiatric hospital. The case is pending because the deadline for its consideration has been extended.

Case Description 11 (C11). Paramedic Protected by the Police (R 9)

Due to the patient's aggressive behavior, the police were called, who filed a motion with the court to punish the perpetrator for active assault on a public official. As a result of several court hearings, the perpetrator was punished, which was not satisfactory for the injured paramedic, as the perpetrator received a suspended sentence.

Conclusions from the Analyzed Cases

Case C10 had tragic consequences not only for the nurse's health (physical and mental) but above all for the life of her unborn child. Unfortunately, although aggressive behavior by patients is condemned, the penalties imposed are disproportionate to the seriousness of the acts committed and give rise to a strong sense of social injustice. And for the medical staff themselves, they are associated with lifelong trauma.

Aggressive physical behavior by patients usually ends with police intervention, which, in accordance with applicable law, prepares documentation and secures the evidence necessary to punish the perpetrator through criminal proceedings. Punishment of patients in Poland is relatively rare, and the penalties imposed are disproportionate to the acts committed and their consequences, and are not satisfactory for the injured employees. In addition, criminal proceedings mean that the victim has to relive the incident each time they visit the court, which leads to growing frustration and discouragement, which can cause depression.

Cases of Aggressive Behavior by Patients Affecting the Functioning of a Medical Facility

Case Description 12(C12). Doctor Asking a Patient to Put on a Mask (R 1)

A patient in the emergency room behaved aggressively after being told that during the SARS-CoV-2 pandemic, it is mandatory to cover the mouth and nose with a medical mask in enclosed spaces. The doctor in charge of the emergency room reported the incident to the Legal Advisors Team, who then submitted a request to the court to punish the perpetrator. At the first court hearing, the defendant stated that he was mentally ill and required psychiatric treatment. In view of the situation, the proceedings were adjourned and the court appointed psychiatric experts to examine the patient. In addition, the patient accused the doctor of destroying his cell phone. However, thanks to the analysis of the incident recorded by the video surveillance system, it was clarified that the incident reported by the perpetrator did not actually take place. The doctor participated in months of hearings, which resulted in the defendant being given a small fine for disturbing the peace and using profanity in the emergency room.

Case Description 13 (C13). The Patient Attacked the Doctor and Destroyed the Equipment (R 2)

A patient under the influence of intoxicants used a powder fire extinguisher to attack a doctor, destroyed medical equipment, and grossly disrupted the proper functioning of the emergency room. This fact was recorded in the medical records along with a description of the situation, and police officers were called to the scene, who drew up the relevant documents and brought charges against the perpetrator. As a result of the court proceedings, the patient was held criminally liable for the offense only in the form of a fine of PLN 500. Which is the equivalent of over 100 EUR and is a very low penalty in relation to the harmfulness of the act committed.

Conclusions from the Analyzed Cases

The obligation to cover the mouth and nose with medical masks resulted from relevant legislation in connection with the pandemic.[6] Despite the order imposed by the regulation in the emergency room, patients did not always comply with the above-mentioned regulations, and when they were warned, they behaved aggressively, accused the staff of destroying their property, and the accused had to prove that the situations presented by the aggressors were fabricated. Patients were usually punished with a fine, the amount of which was grossly low. Employees perceived such situations as unfair and discouraging to work in healthcare. Unfortunately, this points to a visible social problem related to patients' failure to comply with their legal obligations and the difficulty of enforcing this law. ER staff consider aggressive behavior on the part of patients to be a permanent feature of their job. Increasingly, this behavior is not only directed at medical staff but also paralyzes the functioning of the EDs, making it impossible to provide the necessary assistance to those most in need. In addition, such behavior can result in the destruction of equipment and work tools, causing high costs for the medical facility in question.

Table 5.15 summarizes the most important aspects related to aggressive patient behavior in the ED, based on an analysis of the case studies described.

A summary of selected aspects of the analyzed cases related to aggressive patient behavior leads to certain conclusions and observations. Severe forms of aggressive behavior on the part of patients are experienced by doctors, nursing staff, and paramedics alike. The most dangerous form of aggressive behavior, which has severe consequences for ED medical staff, is aggressive behavior in the form of physical assault. In four of the analyzed incidents of aggressive behavior on the part of patients, the police were notified, in one case the hospital's legal team was notified, and in one incident the victim decided to file a private lawsuit. Unfortunately, most incidents of aggressive behavior by patients with particularly severe consequences for employees were not reported to anyone. Most often, the consequence of aggressive behavior by patients was bodily injury requiring the injured employee to take sick leave. In three cases, the consequence was resignation from work. In one case, damage to medical equipment and disruption of work were reported, and in another, unfounded accusations and the need for the victim to participate in court hearings. In one case, there is no information about the consequences of the patient's aggressive behavior.

The cases presented show that patients in the emergency room behave very aggressively, not only verbally but also physically. Physical forms of aggressive behavior by patients have the most severe consequences for medical staff. Unfortunately, these incidents are most often not reported by medical staff. This is mainly due to a feeling of powerlessness and a lack of faith in the effectiveness of such action, as the respondents openly admit. In addition, the negative experiences of medical staff show that this has many unpleasant consequences for the employees themselves who report abuse in this regard. According to the respondents, the current procedures seem to be ineffective and insufficient. There is a lack of real organizational support in such situations, not only in terms of

TABLE 5.15
Summary of Aspects Related to Aggressive Patient Behavior Resulting from the Analyzed Case Studies

	Aspects Analyzed				
Case Study	Employee Staff	Form of Aggressive Behavior	Reaction of the Employee	Reporting Incident	Consequences
Case 1	Nurse	Mental	Private indictment	No	Termination of employment contract
Case 2	Doctor	Mental	None	No	Termination of employment contract
Case 3	Doctor	Mental	None	No	Termination of employment contract
Case 4	Paramedic Medical	Physical	Notification of the police	No	Fractured nasal bone and permanent bodily injury
Case 5	Paramedic Medical	Physical	None	No	Scar on the back and trauma after the incident
Case 6	Nurse	Physical	None	No	Facial injury
Case 7	Paramedic Medical	Physical	None	No	Exposure to blood
Case 8	Physician	Physical	None	No	Sick leave
Case 9	Nurse	Physical	Police notification	No	Miscarriage and sick leave
Case 10	Paramedic Medical	Physical	Police notification	No	No information
Case 11	Doctor	Verbal	Notification of Legal Counsel	Yes	Unfounded accusation, need to participate in court hearings
Case 12	Paramedic Medical	Physical	None	No	Injuries including swelling of the nose
Case 13	Physician	Physical	Police report	Yes	Destruction of medical equipment and disruption of emergency room work

Source: Study based on research.

ensuring the physical safety of medical staff (in the form of procedures and effective security personnel in the workplace) but also in terms of legal and psychological support.

SUMMARY

Incidents of aggressive behavior toward emergency room medical staff by patients occur very often and their effects are particularly harmful. The consequences of such behavior can be considered on many levels. Not only in the context of the medical worker's private and professional life: physical losses, negative impact on the professional development, and health of the employee, on family life, but also in the context of the organization (financial losses, low morale, and employee dissatisfaction) and even society as a whole.

The effects of experiencing aggressive behavior from patients have various health consequences. Health consequences are one of the reasons for exclusion from the labor market, which is a worrying phenomenon in the context of a shortage of medical staff, especially specialists in the field of emergency medicine. Unfortunately, due to the excessive number of formalized procedures aimed at determining the circumstances of the incident and referring the case to court, employees do not report incidents, even those resulting in injury. Given the low penalties and even impunity for patients who behave aggressively, employees feel helpless and unfairly treated. Performing a profession of great importance and responsibility, they bear high emotional and even health costs, without the possibility of protecting themselves against them. This leads to frustration, lack of commitment, and even a desire to leave their jobs.

The high harmfulness of incidents of aggressive patient behavior results in enormous individual, organizational, and social costs. Analysis of the above studies indicates that in order to improve the situation, it is crucial not only to introduce mandatory reporting of incidents and a central register of such events in a given country but also to simplify procedures both within the organization and in court cases. The cases analyzed also point to the need to raise awareness of the rights of medical staff and their protection as public officials, as well as to increase penalties for acts of violence against emergency room medical staff. In the case of violations of the law and acts of aggression, a broad information campaign and the presence of announcements and information, especially in emergency departments, seem necessary.

NOTES

1 Polish law provides protection for doctors, nursing staff, and paramedics in emergency departments, similar to that provided for public officials [Act of December 5, 1996; Act of July 15, 2011; Act of July 25, 2001].
2 This refers to hospitals selected for the study located in Poland.
3 This refers to Poland as well as those countries where these regulations are insufficient.
4 The obligation to wear identification badges in Poland is specified in Article 36 of the Act of April 15, 2011, on medical activities.
5 The "serological window" is the period of time from the moment of infection to the possibility of detecting antibodies produced by the body using methods applied in laboratory diagnostics.
6 In Poland, this obligation resulted from the Regulation of the Council of Ministers of May 6, 2021.

6 Multilevel Practical Strategies for Addressing Management Challenges

The empirical research conducted allowed for the formulation of conclusions and practical recommendations that may prove useful for management practices not only in the context of occupational safety but also in terms of employee management in the healthcare sector. The use of several research methods allowed to explore the problem of aggressive patient behavior and understand how emergency room medical staff deal with patient abuse. The chapter also includes good practices from medical facilities that can inspire decision-makers and healthcare providers in dealing with the problem described. The need for change and a strategic approach to preventing and mitigating the negative effects of violence in the emergency department (ED) was identified, which will allow for the creation of a sustainable working environment for medical staff.

SUMMARY OF RESEARCH RESULTS ON AGGRESSIVE BEHAVIOR OF PATIENTS

Aggressive behavior by patients in healthcare facilities occurs worldwide [di Martino 2003, p. 5–34]. Healthcare workers are particularly exposed to aggressive behavior of an external nature, which results from their relationships with patients and their families.

The results of the study indicate that ED medical staff in Poland are exposed to various forms of aggressive behavior, which is also confirmed by earlier studies [Grudzień et al. 2015, p. 244].

Forms of Aggressive Behavior Exhibited by Patients

The survey indicated that doctors, nursing staff, and paramedics working in EDs most often experienced aggressive behavior from patients in **verbal form**, less often in physical form. Previous literature studies also confirm that verbal aggression is most common toward medical personnel [Fernandes et al. 1999, p. 1247; Le et al. 2021; Rabin et al. 2025]. According to the respondents, verbal aggression from patients is a common phenomenon and employees treat it as an integral part of their work, which has also been confirmed in studies on the exposure of emergency room nursing staff [Lickiewicz et al. 2018, p. 269; Le et al. 2021]. Respondents indicated that aggressive behavior on the part of patients' loved ones, such as family members

DOI: 10.1201/9781003716631-7

and friends, was also significant, which is confirmed by the results of previous studies [Płaszewska-Żywko et al. 2019, p. 145].

Among **the physical forms of aggressive behavior** of patients, waving arms was the most common, although hair pulling, choking, and biting also occurred. Biting by a patient was particularly severe in its consequences. Biting qualifies as exposure to potentially infectious (contagious) material.[1] Aggressive behavior in the form of biting also qualifies as an accident at work, which is regulated by law.[2] Although the incident of being bitten by a patient qualified as occupational exposure to biological agents and met the four conditions given in the definition of an accident at work, it was not reported anywhere.

Intimidation and threats were the most common **forms of psychological** aggression. The most common verbal forms were shouting, swearing, verbal abuse, and verbal altercations. Previous studies have shown that ED employees were frequently exposed to aggressive verbal behavior in the form of raised voices, swearing, verbal skirmishes, ironic comments, and discussions [Szwamel, Sochocka 2014, p. 153; Płaszewska-Żywko 2019, p. 146].

Doctors, nursing staff, and paramedics working in the ED also experienced **indirect** aggressive behavior from patients. Studies have shown that a few respondents received offensive text messages and emails. Most ED medical staff experienced problems related to posts on social media. Periodically, humiliating and unverified information was made public on the Internet.

The study shows that most aggressive behavior from patients in the ED is observed "at night" and "in the evening." This may be related to reduced staffing and limited assistance, which is consistent with the literature [Czyżewski, Rybak 2018, p. 113; Lickiewicz 2018, p. 270].

The research presented in this book showed a weak **correlation** between occupation and physical forms of aggressive behavior. Analysis of the survey results did not reveal any significant differences between the frequency of aggressive behavior experienced by patients and the occupation of medical staff. The professional groups surveyed in EDs in Poland were exposed to physical, psychological, and verbal forms of aggressive behavior from patients with similar frequency, which does not correspond with the results of studies indicating that nursing staff experienced aggressive behavior from patients much more often than doctors [Lickiewicz, Piątek 2014, p. 18; Kowalczuk, Krajewska-Kułak 2017, pp. 114–115].

The analysis of the survey data did not reveal any differences in exposure to aggressive behavior from patients in terms of the gender of medical staff. Women and men were equally exposed to physical, psychological, and verbal forms of aggressive behavior. This refutes the stereotype prevalent in public opinion that women are less likely to experience aggressive behavior from patients because they are perceived as weaker and more delicate. Previous studies have also found no statistically significant differences between the scale of aggressive behavior by patients and the gender of staff [Kowalczuk, Krajewska-Kułak 2017, p. 115]. The results of the study did not confirm the statement that aggressive behavior by patients in physical form is less frequently used against people representing professions with higher social prestige.

Reactions of Medical Staff to Acts of Aggression

When dealing with patients exhibiting aggressive behavior, doctors, nursing staff, and paramedics reacted in different ways. The study showed that regardless of the form of aggressive behavior, the most common reaction of the surveyed employees was to ask for a change in behavior. Emergency room medical staff relatively rarely engaged in discussion with the patient, believing that it would not contribute to a change in their behavior. Half of the respondents asked the police for help, and about 30% of the respondents used the option of forced immobilization. The survey results showed that forced immobilization was a response dependent on physical and psychological forms of aggressive behavior, while calling the police for help was dependent only on physical forms of such behavior, which is consistent with previous studies [Czyżewski, Rybak 2018, p. 113]. Similarly, the results of qualitative research showed that the respondents called the police for help only occasionally, mostly in cases of aggressive behavior in the form of a physical attack that caused damage to property for which they were responsible.

The study showed that as a result of contact with an aggressively behaving patient, medical staff in the ED experienced a range **of negative emotions**. A large number of respondents felt anger, rage, agitation, helplessness, anxiety, and fear. Many people felt uncertainty, and some felt sadness, regret, and despair. Anxiety and fear were emotions that arose as a result of contact with patients exhibiting physical and psychological forms of aggressive behavior. Anger most often arose as a result of contact with patients exhibiting psychological forms of aggressive behavior, and rage most often arose as a result of contact with patients exhibiting verbal forms of such behavior. The emergence of these emotions can affect the functioning of employees and, as a consequence, lead to work disorganization, medical errors, a decrease in work efficiency, and a reduction in the quality of services provided, which is similar to the results of previous study [Szwamel, Sochocka 2014, p. 153].

The results of the survey indicate a correlation between **the functioning of registers of aggressive patient behavior** and the province in which the ward is located. In turn, the results of the qualitative study indicate that registers of aggressive patient behavior do not exist. Despite the large scale of aggressive behavior on the part of patients, the surveyed ED employees did not report these incidents. Incidents related to aggressive patient behavior were only recorded by ED staff in electronic medical records kept in IT systems. The study indicates that ED medical staff did not report incidents of aggressive behavior on the part of patients due to the lack of legal regulations in this area. There are no regulations that would require medical facilities to keep records of aggressive patient behavior, nor are there any regulations that would impose such an obligation on victims, in this case ED medical staff. Employees sporadically reported cases of aggressive patient behavior, and only those in the form of physical attacks that would result in consequences, as shown by the analysis of cases with particularly severe consequences for employees. The documentation of incidents of aggressive patient behavior in physical form is confirmed by other studies [Burak et al. 2021, p. 114]. The lack of reports and records of aggressive behavior on the part of patients makes it impossible to provide the actual number of such incidents. Reporting incidents of aggressive patient behavior is a broader problem, as shown by the research of Markiewicz, who states in her work that nursing staff do not report incidents of aggressive patient behavior because they consider reporting unnecessary even when the injury lasted longer than 7 days and was associated with pain at the site of the impact [2012, p. 95]. Similar conclusions are presented by Alharthy et al., who point out that employees do not report incidents of aggressive patient behavior because they consider reporting to be useless and, in addition, do not know where to report them or fear negative consequences [2017, p. 28]. Unfortunately, qualitative research conducted in this regard has confirmed this.

Research by Kaeser et al. has shown that employees are reluctant to report incidents of aggressive behavior by patients [2018, p. 4] and that this is a significant global problem [Richardson et al. 2019, p. 272]. What is more, often even the victim is not interviewed [Sachdeva 2019, Davids et al. 2021; Rabin 2025], let alone analyzing cases to effectively prevent similar incidents in the future. It is therefore not surprising that medical staff feel and speak out about the lack of support in this area [Tarczoń et al. 2025]. Research conducted in the United Kingdom has shown that among 96% of young doctors who have experienced patient aggression in the workplace, none of them filed a police report or received subsequent support.

Moreover, research conducted in Poland has shown that employees perceive a lack of interest and support from their employers in reporting incidents involving aggressive behavior by patients in the ED, which is also confirmed by the literature on the subject [Gacki-Smith et al. 2009, p. 340; Aladwan, Dalahmeh 2022, p. 340].

Surveys have shown that in the first phase of the SARS-CoV-2 pandemic, the number of aggressive behaviors by patients decreased, which may be the result of restrictions on accompanying persons in hospitals, especially in the ED. However, during this time, there were cases of aggressive behavior by patients as a result of requests by ED staff to wear protective masks. As subsequent phases of the pandemic emerged, the number of incidents related to aggressive patient behavior increased, as confirmed by research conducted in the form of expert interviews.

The results regarding the identification of risks related to aggressive patient behavior in the ED in the occupational risk assessment forms of doctors, nursing staff, and paramedics are ambiguous. More than half of the doctors and paramedics surveyed reported that occupational risk assessment forms included risks related to aggressive patient behavior, unlike nursing staff, whose forms did not include such risks. Only in some medical facilities do the occupational risk assessment cards for ED medical staff include aggressive patient behavior. Based on the content of the occupational risk assessment cards for ED medical staff, it can be concluded that employers are concerned with fulfilling their legal obligations in the field of occupational health and safety, and carry out "occupational risk assessments" which take into account the threat of aggressive behavior, but not in all analyzed departments and not for all jobs within the same ED. The occupational risk assessment forms lack information on specific forms of aggressive behavior as well as estimates of the risks associated with such behavior.

Causes of Aggressive Behavior in Patients

Among the most common **causes** of aggressive behavior of patients in the ED, respondents indicated the intoxication of patients, whose percentage in the ED was significant, as confirmed by earlier reports [Roche et al. 2010, pp. 13–22; Grudzień et al. 2015, pp. 244 and 39; Czyżewski, Rybak 2018, pp. 112–113; Szkup et al. 2018, p. 120]. Another reason for the escalation of aggressive behavior in patients was the use of psychoactive (intoxicating) substances other than alcohol. Pain and suffering caused by illness, difficult life circumstances, and a lack of security were relatively rare causes of aggressive behavior, which is consistent with previous studies [Frydrysiak et al. 2014, p. 389; Szwamel, Sochocka 2014, p. 152; Czyżewski, Rybak 2018, p. 113]. In turn, the results of qualitative research indicate that nicotine craving was the cause of aggressive behavior among patients in the emergency room. The reasons that seem particularly important are the insufficient amount of time that staff devote to explaining the situation and provocative behavior on the part of medical staff, which is most often noticed by patients with mental disorders. Another important cause was a lack of understanding of the situation in which patients find themselves, downplaying their complaints, and negligence in treatment and care, which is consistent with previous studies [Frydrysiak et al. 2014, p. 388]. The results of our own research showed that the causes of aggressive behavior in patients in the EDs studied had not been analyzed to date. This indicates that the management of the facilities studied did not take any measures to improve the safety of medical staff in the ED in the context of aggressive patient behavior.

Consequences of Aggressive Patient Behavior

Aggressive behavior of patients in EDs causes a number of **negative individual and organizational effects**. The individual effects of aggressive behavior of patients in EDs most often include dissatisfaction with work, loss of dignity and professional pride, prolonged stress, and general health problems.

Organizational effects include disruption of work, high risk of medical errors and reduced quality of services, as well as decreased work efficiency, which can lead to longer waiting times for admission by a doctor. Case studies indicate that aggressive patient behavior can have severe consequences for staff, such as the need to change jobs or permanent damage to health. The effects of aggressive patient behavior in the ED have been classified and summarized in Table 6.1.

The individual effects of aggressive patient behavior have negative consequences for the organization. General health problems, prolonged stress, and depression lead to staff absenteeism, which translates into high turnover and even staff turnover, decreased work efficiency, and additional costs.

The survey results showed that the most commonly used methods of coping after contact with an aggressively behaving patient were playing sports, listening to music, and meeting with friends. Meeting friends is particularly important in the case of difficult experiences due to the possibility of obtaining support. Similar results were obtained by Szwamel and Sochocka [2014, p. 152]. Respondents also indicated the use of psychoactive substances, mainly alcohol.

TABLE 6.1
Effects of Aggressive Patient Behavior in the Emergency Departments

	Effects of Aggressive Patient Behavior
Individual (biological, cognitive, emotional)	Biological • Long-term stress • General health problems • Anxiety and depression Cognitive • Loss of dignity and professional pride • Job dissatisfaction • Low self-esteem • Resignation from the profession • Feeling of powerlessness and lack of influence on patient education Emotional • Discomfort • Frustration
Organizational	• Disorganization of work • Risk of medical error • Staff absenteeism • Decrease in work efficiency • Callousness toward patients and lack of compassion • Decrease in the quality of services provided • Costs associated with employee sick leave • High turnover of nursing staff and doctors • Costs of renovations and purchase of new equipment • Longer waiting times for appointments with doctors

Source: Study based on research results.

The medical facilities analyzed lack procedures for dealing with threats related to aggressive patient behavior. There is a belief that analyzing incidents involving aggressively behaving patients is irrelevant and that preventive measures are ineffective. The research results obtained, in conjunction with an earlier analysis of the literature on safety culture, allow us to conclude that there is a so-called **weak work safety culture** in the analyzed EDs. Although aggressive behavior by patients in the ED is not justified, psychological assistance is not provided to employees who have been subjected to aggressive behavior by patients, which is consistent with previous studies [Rabin et al. 2025]. This may indicate a lack of genuine interest in occupational safety issues as well as organizational support from the management of medical facilities in creating a culture of occupational safety.

When it comes to increasing employee safety, employees point to the need for training in dealing with aggressive patients, handling agitated patients, interpersonal communication, applicable law, and self-defense techniques, which is also consistent with previous studies [Lickiewicz, Piątek 2014, p. 17]. Training is an integral part of the strategy of creating a sustainable and supportive environment for employees and patients. It helps employees increase their sense of safety by developing skills for dealing with aggressive patients [Szwamel, Sochocka 2014, p. 152; Heckemann et al. 2015, p. 25].

Most of the surveyed ED medical staff did not use the legal protection provided for public officials, which is consistent with earlier research results [Szwamel, Sochocka 2014, p. 153; Mikos et al. 2018, p. 59]. The lack of action in this area is due to overly complicated procedures that require a great deal of commitment, financial outlay, and time spent traveling to state institutions. The vast majority of the surveyed employees were in favor of establishing entities whose role would be to educate patients about the legal protection available to medical personnel. They believe that radical measures are needed to limit the escalation of aggressive behavior by patients.

STRATEGIES AT THE GOVERNMENT AND REGIONAL LEVELS

When analyzing possible practices, the literature points to individual measures implemented by entities as well as comprehensive strategies that combine measures in several areas. Following are examples of good practices implemented at the government and regional levels to protect medical staff from aggressive patient behavior, which may inspire others to take action. The examples discussed are taken from an analysis of the websites of individual institutions.

The Case Study of the National Health System in the United Kingdom

The National Health Service (NHS)[3] brings together three of the four publicly funded healthcare systems in England. Efforts to improve the safety of healthcare workers began with a change in awareness of occupational safety and the development of an appropriate **occupational safety culture**. This required strong commitment from senior management, which is one of the main factors for the successful implementation of any change. In cooperation with other social partners, including trade unions, appropriate procedures were developed and then implemented in individual units [NHS 2020]. The standards developed were based on risk assessment, supporting the creation of a sustainable working environment and protecting NHS staff from abuse and aggression. Furthermore, to help service providers share and communicate the risks of violence and the mechanisms for controlling it, specific guidelines were developed for the NHS standard on preventing and reducing violence. The document briefly describes detailed guidelines and actions that should be taken by cooperating entities to identify, assess, and counteract risks in the workplace. In addition, it briefly outlines ways to change employee behavior to support a culture of safety within the organization.

One important way to implement the program is through intensive education and training of NHS staff, which will help change staff behaviors and habits. In addition, such training will develop employees' skills to prevent and respond appropriately to incidents of aggression from patients. At the same time, NHS England, in collaboration with the Royal Society for Public Health (RSPH), has developed an e-learning course entitled "Introduction to Preventing and Reducing Violence." This course was developed based on the identified training needs of staff and was piloted among staff across England, who rated it as "useful."[4]

The Case Study of the National Institute for Occupational Safety and Health

The National Institute for Occupational Safety and Health (NIOSH)[5] in the United States conducts research and develops recommendations for preventing work-related illnesses and injuries. NIOSH works with industry, labor organizations, and universities to improve occupational safety and health throughout the United States. On its website, it encourages the development of appropriate strategies to counter violence in the workplace and the implementation of measures at the organizational level, depending on the specific nature of the organization. One of the measures taken by NIOSH is to raise awareness among hospital employees and managers about creating a healthy and safe work environment through an information campaign. To this end, it publishes brochures and research reports containing guidelines for both individual employees and the hospital as a whole. The proposed brief recommendations concern how to design a safe working environment, focusing on both the physical and non-physical working environment. In terms of the physical working environment, it focuses, among other things, on monitoring, protection, assistance in patient waiting rooms, and appropriate interior design to minimize the use of physical aggression. On the other hand, when talking about the non-material work environment, it points to the need for appropriate administrative control and modification of employee behavior through appropriate training. In addition, it suggests how to deal with emerging patient aggression and how to tone down manifestations of aggressive patient behavior. This may be an excellent example of so-called checklists for many directors of such facilities and HR departments, supporting their work and creating a safe work environment.

According to NIOSH, at the organizational level, all hospitals should develop a comprehensive **violence prevention program**. There is no universal strategy for preventing violence, as risk factors vary from hospital to hospital and department to department. NIOSH recommends that hospitals establish interdisciplinary committees, including direct care staff and union representatives, to identify risk factors in specific work situations and jointly develop strategies to reduce them.

Further recommendations concern the individual level, where all hospital employees should remain vigilant and cautious when interacting with patients and visitors. They should actively participate in safety training programs and be familiar with their employers' policies, procedures, and materials on violence prevention. For example, a short brochure published by NIOSH, dedicated to hospital staff, identifies the causes of aggression, potentially dangerous situations, and provides tips on how to behave in such situations.

The Case Study of the Massachusetts Health and Hospital Association

In the United States, there are also relevant regulations at the regional level – in individual states – aimed at protecting employees and creating a safe and sustainable work environment. For example, in Massachusetts, the Massachusetts Health and Hospital Association (MHA) has developed documents to support hospitals in creating policies for preventing workplace violence.[6]

The MHA safety guidelines provide key steps for creating a safe working environment in hospitals. The documents standardize the reporting of incidents in healthcare facilities in Massachusetts. The document defines relevant phenomena and behaviors, points to related legislation, outlines key steps in preventing violence in hospitals, and identifies sources of risk, including selected issues related to the physical work environment that support workplace safety.

The key elements of the documentation are, of course, procedures and actions such as patient screening, incident reporting and classification, and post-incident investigation. The MHA also recommends appropriate IT systems to support the reporting of such incidents (e.g., Perspective, RL Solutions, Safe Spot, Quantros, Omnigo Software, Guardtek, Safety Event Reporting System (SERS), QED STARRS EMR, and Midas). Finally, MHA identifies necessary training areas for hospital staff, which are an important element in improving hospital safety.

STRATEGIES AT THE ORGANIZATIONAL LEVEL

The problem of aggressive patient behavior toward medical staff occurs in many countries. When planning the implementation of interventions, it is worth learning about solutions that have proven effective in other facilities. There are good practices around the world aimed at eliminating cases of aggressive behavior by patients. In Australia, there are special regulations in place to protect hospital staff from socially unacceptable behavior by patients. For individuals who threaten or physically assault medical staff, Australian law provides for penalties of 5 to 10 years' imprisonment. Mental illness does not exempt these individuals from being questioned and charged with a crime.

The Case Study of the Royal Melbourne Hospital

The Royal Melbourne Hospital makes every effort to provide patients with the highest level of care and to ensure the safety of the staff providing that care. The hospital management actively supports measures to prevent violence and aggressive behavior by creating a work environment that promotes hazard identification, risk assessment and control, training and education, and the reporting and effective management of all incidents. The hospital has established a procedure for safe and positive behavior in the workplace and a code of conduct that includes standards of behavior that employees, patients, their families, and visitors are required to follow. The hospital has an **Aggression and Violence**

Prevention Committee that provides leadership in preventing and managing violence and aggression. The committee develops procedures related to violence, analyzes incidents, and makes recommendations after incidents occur. Employees are encouraged to report all cases of violence through the IT system. A culture of safety that does not tolerate violence is promoted, and all incidents of violence are monitored. Training is provided for staff on how to deal with violence, and support is given to staff who have suffered trauma or illness as a result of violence. The hospital has a **"trauma counseling" program** for all employees exposed to violence. Employees are required to follow the procedure, participate in training on managing violence in the workplace, and report all incidents and potential cases of violence through the IT system. In circumstances where a violent incident has occurred, employees have the right to file an official report with the police. Employees do not need to ask their employer for permission to file a report. However, before filing a report, it is recommended that they notify their supervisors and provide all relevant information so that the hospital can provide support to the employee. To report an incident, it is sufficient to provide name, contact details, parties involved in the incident, details of the nature of the incident, and the level of injury. An interesting initiative involves sending **letters to patients** who have behaved aggressively in the hospital. The letters state that if the patient returns to the hospital, they will have to commit to behaving appropriately, and this warning is noted in the patient's medical records. In addition, weapons and objects that could cause harm, as well as drugs and alcohol, are prohibited on hospital premises. Hospital security has the authority to search any person and, if illegal substances are found, has the right to confiscate and destroy them. If weapons are found, security may immediately confiscate them and report the incident to the police [*The Royal Melbourne Hospital*[7]].

Behavioral Assessment Rooms

Another effective intervention taken by the Department of Health and Human Services of the State of Victoria in Australia to combat aggressive behavior by patients in hospital EDs are the so-called behavioral assessment rooms. Patients in hospital EDs who may harm themselves or others and who are agitated and behaving aggressively are placed in so-called behavioral assessment rooms.

Behavioral assessment rooms are a separate space that ensures privacy and where an interdisciplinary team works. The team's work in behavioral assessment rooms focuses on activities aimed at reducing the risk of injury to patients and staff, ensuring a positive experience for patients, and reducing stress for other patients and visitors to the ED. The Department of Health and Human Services has developed guidelines to help establish appropriate procedures and policies for the use and design of behavioral assessment rooms. The guidelines apply to both public and private hospitals that have or plan to introduce a behavioral assessment room in their ED. Behavioral assessment rooms are not standard in hospital EDs, mainly because there are not enough staff to ensure the safety of staff and patients or to support patient care in the department [*Department of Health Victoria*[8]].

Use of a Checklist

Australian mental health centers such as the Royal Brisbane and Women's Hospital, Prince Charles Hospital, Ipswich Hospital, Caboolture Hospital, Gold Coast Hospital, and The Park – Mental Health Center have introduced **the Safewards program**. The program was developed in the United Kingdom by Professor Len Bowers and involves a departure from traditional education-based initiatives, proposing an alternative approach to preventing violence and improving safety in the workplace. This initiative involves the use of a "checklist," which is a tool for predicting the escalation of violence by patients within the next 24 hours. The list includes six variables that relate to the risk of violence. These are confusion, irritability, loudness, physical threats, verbal threats, and attacking objects. The violence checklist is a reliable risk assessment tool that does not require a lot of time [*Occupational Violence Prevention…* 2016, pp. 27–28]. It helps employees predict certain patient behaviors and take actions that will reduce the likelihood or prevent aggressive behavior by the patient.

Employee Mutual Support Program

The Mental Health Center in Park has also introduced a mutual support program as part of a multifaceted strategy to reduce the incidence and negative effects of patient violence against medical staff. The mutual support program is available to all employees who have experienced violence from patients at the facility. The project is run by a small team of volunteers who have participated in a training program on mutual support principles and stress reduction techniques. As part of the program, an employee who has experienced violence from a patient is monitored at least three times: on the day of the incident, for the next three days, and ten days later. During the last visit, an assessment is made as to whether, the person needs professional support under the program [*Occupational Violence Prevention…* 2016, p. 56]. The requirement to support employees who have experienced violence during their employment is a legal obligation. Therefore, after each incident, the program provides victims who have experienced violence with prompt and effective support for their physical, emotional, and mental well-being. Support programs have been found to benefit victims of workplace violence, help reduce the number of incidents, decrease their frequency, support the employee's return to work, and alleviate symptoms associated with *post-traumatic stress disorder* (PTSD*)* [*Occupational Violence Prevention…* 2016, p. 57].

Building Employee Solidarity – the Code Grey Initiative

Meanwhile, the ED in Townsville is testing the so-called *Code Grey* model, which was developed due to the ineffectiveness of existing communication strategies aimed at de-escalating violence. This initiative involves taking measures to ensure a safe clinical environment for patients. The essence of "*Code Grey*" is close cooperation between trained medical personnel and security staff. *Code Grey* assumes that if workplace violence is seen as a clinical problem and the response to incidents is based on clinical principles, a team approach can help overcome the fear and vulnerability of employees experiencing violence. The initiative is accompanied by the message "we are all in this together," which has a measurable effect. The introduction of "*Code Grey*" is a positive initiative in the de-escalation of violence throughout healthcare, but particularly in EDs. In 2014, the Australian Department of Health and Human Services published the "*Code Grey*," which sets standards and best practices for responding to incidents of violence and aggressive behavior by patients in healthcare. According to these provisions, "*Code Grey*" is defined as a coordinated response by all hospital staff to actual or potential aggressive behavior or violence that does not require the use of weapons [*Occupational Violence Prevention*... 2016, pp. 28 and 51–52].

The examples given mainly concern developed countries, where many government institutions are strongly involved in combating violence and aggressive behavior by patients toward hospital staff. Many significant initiatives are being developed to improve the safety of medical workers. One of the key measures in this direction is, among others, the creation of a simplified mechanism for staff to report cases of violence and forcing medical facilities to publicly report incidents involving violence and aggressive behavior by patients [*Emergency Care in Victoria*[9]].

The initiatives presented show the measures taken to de-escalate aggressive behavior by patients in medical facilities around the world. The strategy to reduce violence and aggressive behavior by patients in hospitals is based primarily on changing the organizational culture and covers topics related to prevention, early intervention, and post-incident response. This change is not easy to implement in healthcare facilities and also requires adequate preparation.

IMPLEMENTING CHANGE IN THE HEALTHCARE SECTOR

Preventing patient aggression toward staff requires a systematic approach, including changing the attitudes of hospital staff. The results of the study showed not only the enormous scale of this phenomenon but also the failure of affected employees to report such incidents and the phenomenon of so-called organizational silence. This is very disturbing, so in order to create a sustainable work environment, but also to eliminate such behavior, it seems necessary to change not only the procedures but also the attitudes of the employees themselves. These are therefore measures that an employer can take to eliminate aggressive behavior and create a culture of safety at the organizational, team, and individual levels.

Changing Staff Attitudes at the Organizational Level

When starting to change attitudes within an organization, it is worth asking a few key questions:

- What kind of change is this? Should this be a participatory or directive change?
- What is the dilemma and why is it important?

This distinction means radically different behaviors on the part of the change manager [Hersey, Blanchard, Johnson 2000, p. 391].

In **the participatory approach**, the emphasis will be on employees to "do it for the manager" or "do it for themselves" and understand why reporting aggressive behavior is so important. For example, workshops can be organized where participants acquire the necessary knowledge and learn new behaviors through simulation. Then, the behavior of the individual – the healthcare worker – will change in a way that is desirable from the point of view of behavioral change, which after a few weeks may turn into habits. The behavior of individual employees (especially so-called opinion leaders) will change the functioning of the entire team. This type of organizational change can be understood as an unforced change from the bottom up of organizational structures.

The opposite is **a change referred to as directive**, which is rather a change of a professional nature, running from the top down through organizational structures. The leader uses their formal authority, resulting from their position in the organizational structure, to create a new procedural framework for the functioning of employees. New procedures should be introduced, along with a system of incentives and controls for reporting. Such regulations will influence medical teams and, consequently, individual employees. Over time, employees will accept the new situation (often after a period of resistance or testing the consequences of implementing the new system) and their attitudes will change. Acceptance will create a need for knowledge about new rules for maintaining safety at work in the face of aggressive patient behavior, which in turn will reinforce the new attitudes of the staff.

Both models seem to be scenarios for change that can be successfully implemented in the healthcare sector. Effective change will occur when a supervisor imposes a new framework of behavior resulting, for example, from modifications to statutory regulations. This also applies to situations where staff begin to behave differently because they understand the usefulness of the new approach and have had the opportunity to successfully practice new behaviors in a safe environment. Before making a final choice, it is worth considering organizational factors such as [Rosiński, Filipkowska 2009]:

- Time needed to implement the change (directive change is shorter, participatory change is longer),
- Ways of involving staff (in directive change, involvement is minimal, while in participatory change, involvement is key to the success of the process),
- Budget (directive change carried out efficiently is cheap, but its costs increase over time; the cost-effectiveness of participatory change is the opposite) – consultations and workshops conducted in a short period of time are expensive, the longer the change, the lower the costs, as many activities can be carried out within the organization without paying for external services.

When choosing the final strategy, it is worth considering not only organizational guidelines but also processes at the level of individual employees. As Muraven M.,

Gagné M., and Rosman H. [2008], successful organizational changes are based on encouraging employees to make independent choices, which is associated with a pleasant incentive. Such an approach results in more lasting and less burdensome habit formation than in situations where changes are imposed on employees and they feel treated as objects. Analyzing previous publications and studies, it seems that when changing procedures and employee behavior toward patient aggression, it is worth using a participatory approach to managing change, working together with employees in a process of dialogue to develop ways of operating in a specific organization.

The second organizational aspect related to lasting change in attitudes is a dilemma that can be boiled down to the question: where to start the change? A common answer may be to involve employees and those directly managing teams in the change process, while excluding senior management (i.e., those managing teams of managers). Despite its frequency, this approach seems to lead to failure, as the following scenarios often occur:

- project failure due to the ambiguous attitude of senior management,
- management functioning in a situation where they implement change in their teams and at the same time play their previous roles in their relationships with their superiors.

A lasting change in attitudes within an organization should begin with a change in the attitudes of managers – they are the ones who organize the work environment in which new behavioral habits can emerge, eventually becoming skills (Figure 6.1).

The right decision regarding the nature of change implementation (participatory/directive) and the creation of an environment of change for employees by senior management are key aspects for controlling attitude change at the

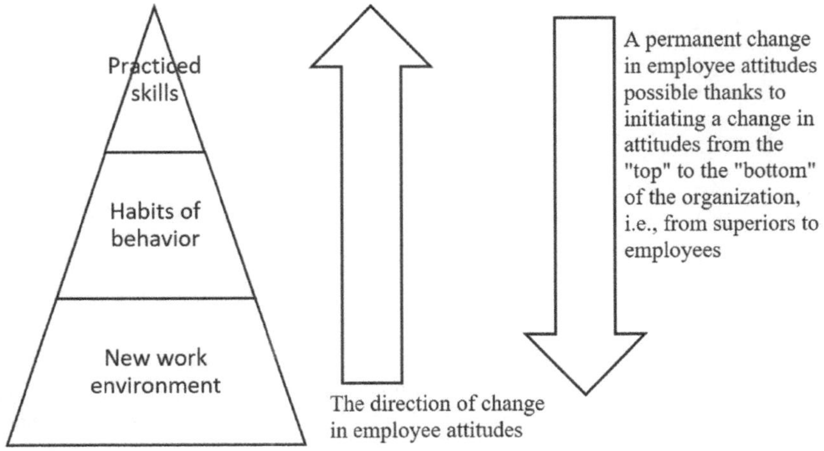

FIGURE 6.1 Proposed direction for implementing organizational change

organizational level. Beneficial actions in these two areas neutralize a significant part of the causes of failure in implementing change [Blanchard K. 2013, p. 168]. At the same time, they are necessary but not sufficient conditions, as aspects related to the employee team and individual employees must also be taken into account.

Typical Concerns That Arise when Implementing a New Course of Action

The process of attitude change described by Kurt Lewin [1947] can be applied in three stages: unfreezing, change, and refreezing, also to the management of a team of employees in an organization undergoing transformation [Dyląg A., Rosiński, 2008, p. 122]. This process is associated with specific, predictable concerns that arise among a group of employees. The individual fears and their duration are described in Figure 6.2. The list of individual fears and their order are based on the concept of K. Blanchard [2013, pp. 169–184], but the specific scope of each fear differs from the concept mentioned earlier. This step results from bringing the model closer to the reality of change implementation, unfortunately with a loss in the symmetry and elegance of the original model (Figure 6.2).

The best approach for a change manager is to use **an instructional style** (according to K. Blanchard's categorization), i.e., clearly informing employees how the change will take place, what their supervisor will expect in the near future, what the next steps will be, what we expect from each individual, and how their functioning will change in the new situation. In this case, using an instructional style

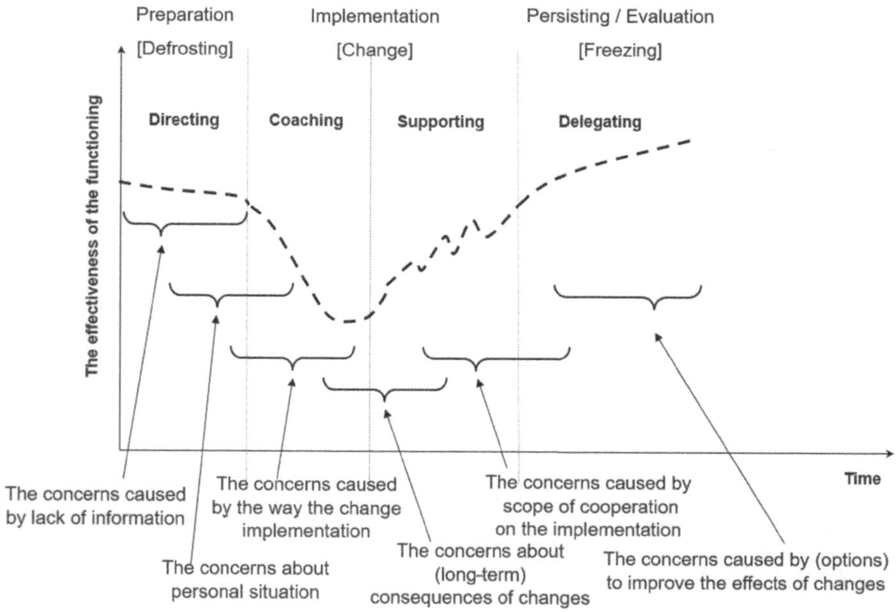

FIGURE 6.2 Employee concerns in the change process

is also a preventive measure – it reduces the amount of gossip. The preparation stage of change causes a lot of anxiety, so if we do not clearly inform employees about what will happen, the area of uncertainty will be filled with gossip – this is a natural reaction when there is anxiety and insufficient information. While at the stage of preparing for change we are dealing with a certain stagnation in team effectiveness, the launch of activities related to the implementation of change can cause a sharp decline in the effectiveness of the group of employees. Concerns arise about how the change will be implemented and about the long-term consequences of introducing the change. A leader who wants to manage change effectively should change their communication style from instruction to consultation. This means facing reluctance and protests, learning more about the situation – the reasons for concern – and asking the team about possible actions. The final decision on how to proceed should be made by the leader, as the group has not yet accepted the new situation at this stage and is checking whether it is possible to return to old habits and ways of working. The initial stage of implementing change is not easy emotionally for the person leading the change. However, listening to employees at this stage of change implementation creates a pool of ideas and good practices that can be converted into effective organizational procedures. Therefore, despite attacks on the leader/change/organization, it is worth meeting with employees and their reluctance. Paradoxically, the resistance presented here is good "fuel" to drive the entire change – the group resists, but this indicates emotional involvement, which, thanks to the use of a consultative style, can serve the implemented project well.

Although implementation is a single stage, it has two distinct phases: the first (discussed earlier) is related to the start of implementation, and the second refers to the implementation of change over a longer period of time. In the second phase of change implementation, employees may already be tired of the prolonged state of mobilization, but they have usually accepted the change as inevitable and permanent; they are usually convinced that the change makes sense. At this time, concerns about the long-term consequences of the change and concerns about the extent of cooperation in implementing the change ("Who else can we involve?" and "How can we encourage others to get involved in our change?"). Both types of concerns can be reduced at the organizational level by presenting examples of employees who have succeeded and owe their success to the change. Their unequivocal support for the change has a positive impact on the team. At the level of the relationship between the leader and the group, it is worth changing the communication style to a supportive one. The group is tired, but their competence in dealing with the new situation has increased, and they have also accepted the change as their solution. This means that the leader should listen to the team and ask how employees understand the problem and what possible solutions they see. However, specific to the supportive style, the responsibility for developing the final solution remains with the team (in the consultative style, the leader made the final decision).

Consolidating change raises concerns about the possibilities for improving the effects of change. This unusually named concern relates to two areas: concern about the next stages of action ("What will happen next, after this change?") and concern about the re-formalization of the organization. In this case, these concerns are alleviated by delegation as a method of management. It allows employees to take control of the performance of tasks, enabling them to implement their own ideas on how to achieve goals.

Change in Staff Attitudes – Individual Employee Level

The processes **of permanent change in employee attitudes** are phenomena that take place at the level of the individual. It is worth noting here the dynamics of individual reactions to change. For the person managing the change, it is important that employees do not immediately accept the new situation. Even if the change is beneficial, any attempt to eliminate an old habit will provoke resistance and mobilization aimed at restoring the old state of affairs. Only when it becomes clear that a return to the old ways is impossible does acceptance and adaptation to the changed realities of the workplace occur. For the supervisor, this dynamic reaction means that for the change to be successful, the implementation of new behaviors requires at least several weeks of increased monitoring of employees in the area of behaviors that are to change. Otherwise, the subordinates' mobilization aimed at "returning to the old ways" will prove successful, and after a few weeks it will turn out that nothing has changed. Unfortunately, it will be more difficult to introduce another change in such a team.

It is worth noting that this mobilization of energy to protect the status quo, the so-called resistance to change, is often perceived as a threat. However, resistance is not a destructive phenomenon. The contemporary approach to change management requires resistance to be treated as a typical or completely natural reaction, beneficial – under certain conditions – for the effects of change [Dyląg A., Rosiński J. 2008, pp. 117–121]. We need this mobilization to overcome old ways of doing things and persevere in the process of learning new ones.

Paradoxically, resistance intensifies in situations where attempts are made to keep secrets, conceal information about changes, threaten the interests of various groups, or rely on fear as a strategy for introducing change rather than focusing on cooperation. Therefore, instead of pointing out that resistance is bad, it is worth talking to the employee about the reasons for their reluctance. It is worth "leaning in" to the reasons for their reluctance, because, as Kurt Lewin [1947] points out in his classic concept of force field analysis, without neutralizing resistance (inhibiting forces), it may turn out that further incentives will only intensify resistance and employees' attitudes will not change: stronger incentives will be met with proportionally increasing resistance to change (Figure 6.3).

Changing attitudes is not an easy task, especially in the face of already established ineffective behaviors. However, controlling change at the three levels described earlier – organization, team, and individual employee – allows for effective modification of employee behavior in the long term. In the case of shaping the appropriate attitudes of medical staff, such comprehensive action at all levels can contribute to the long-term development of a culture of safety, supporting the creation of a sustainable work environment.

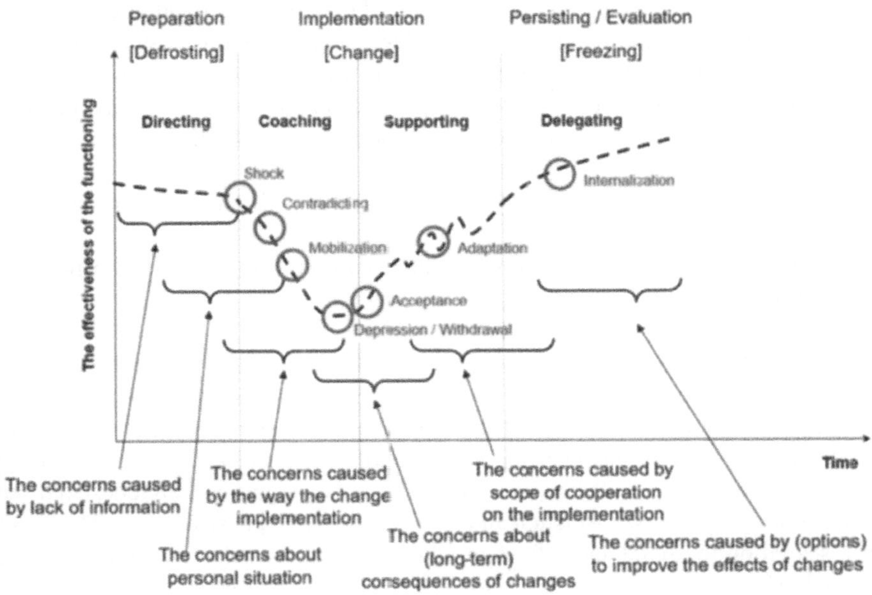

FIGURE 6.3 Dynamics of individual response to change

RESEARCH CONCLUSIONS AND RECOMMENDATIONS

Conclusions from the Research

The aim of the research was to identify factors affecting work safety in hospital EDs related to aggressive patient behavior and to outline recommendations for developing procedures to increase work safety in this area. The results collected at all stages of the research process helped to formulate the following conclusions:

- The scale of aggressive behavior by patients toward medical staff in EDs is high, but the lack of reports and records makes it impossible to give the actual number of such incidents.
- Aggressive behavior by patients toward medical staff in the ED is most often expressed verbally.
- No correlation was found between the profession and the frequency of experiencing aggressive behavior from patients.
- Aggressive behavior by patients is more common at night or in the evening.
- Aggressive behavior toward ED medical staff is most often exhibited by patients under the influence of alcohol or other psychoactive substances, or those who are addicted to these substances.
- Hospital EDs in Poland lack records of aggressive patient behavior, and incidents are voluntarily recorded by staff in the Electronic Medical Records (EDM).

- Despite the increased contact between ED medical staff and aggressively behaving patients, there is a lack of identification and assessment of this risk in occupational risk assessment cards. In 14.3% of the analyzed occupational risk assessment cards of ED staff, there is no identification of the risk associated with aggressive patient behavior, and in 42.9%, there is no assessment of the risk associated with such behavior.
- There is a lack of education on how to deal with patients who behave aggressively.
- ED medical staff occasionally use the legal protection provided for public officials.

The results of the study confirm that aggressive patient behavior in the ED poses a significant threat to the work of medical personnel in this department. It hinders the performance of professional duties and creates a risk of medical error, which can have negative consequences for patients in the form of infection, bodily injury, loss of health, and even death. For employees, it is a significant psychosocial burden, resulting in job dissatisfaction and psychological and physical harm. The results of research on the scale and effects of aggressive patient behavior, combined with the lack of procedures, support, employee education, including specialized training, records of aggressive patient behavior, and failure to take advantage of the protection available to public officials, indicate that it may be crucial to implement an occupational health and safety management system (in accordance with ISO 45001: 2024) and to create a culture of occupational safety.

The study has certain limitations. Due to the SARS-CoV-2 pandemic, many directors of medical facilities did not consent to the study. Therefore, it would be advisable to conduct similar studies in all hospital EDs, commissioned by the Ministry of Health, which would help to increase the research sample and raise awareness among the relevant entities of the need to create a culture of occupational safety in hospital EDs.

Recommendations for Management Practice

The results of the research and the conclusions from the analysis allowed for the formulation of practical recommendations. The measures taken so far to reduce the escalation of aggressive behavior by patients toward doctors, nursing staff, and paramedics in the ED have not brought the expected results. Working in the ED still poses many risks that can harm the life and health of employees. Therefore, it is necessary to introduce systemic changes that will improve the functioning of the ED and have a positive impact on work safety. The first change could be a change in patients' rights to receive medical assistance in the ED. Medical assistance should only be provided to patients brought in by the emergency medical services who are in a state of immediate danger to their health or life.

Due to the scale of aggressive behavior by patients toward medical professionals in emergency departments, there is an urgent need for legislative work and the development of regulations that would impose an obligation on healthcare entities to take measures in the area of occupational safety management. Research shows **that**

doctors, nursing staff, and paramedics expect such measures to be taken. One of the most important changes is the need to introduce into generally applicable **law a provision requiring** the reporting of an incident to the immediate superior, the director of the facility, the police, the securing of evidence of the incident, the preparation of an official note, and the collection of written statements from witnesses to the incident. At the same time, it is recommended **to simplify the procedures** for exercising the status of a public official and to introduce an IT system in which all the information necessary to report an incident and bring the perpetrator to criminal responsibility could be entered quickly and easily.

The issue of breaching professional confidentiality by reporting a patient's aggressive behavior is highly controversial. According to the law, medical professionals are required to keep information about patients confidential, in particular their health status, but this obligation does not apply if it could pose a danger to the patient or others. Aggressive behavior by a patient in the ED may result from their health condition and at the same time pose a threat to staff and other patients, which, in the authors' opinion, entitles the doctor to include information about the patient's behavior in their medical records. Aggressive behavior on the part of a patient may hinder or even prevent the correct diagnosis of the patient and should, in such a situation, be reflected in the medical records [Simińska 2016]. It seems particularly important to oblige ED medical staff to report incidents of aggressive patient behavior in order to make them aware that such behavior is punishable by law. Every case of aggressive patient behavior that results in even minor injury to an employee should be reported to the health and safety department and documented as an "accident at work." It would be useful to create **a central register of incidents** involving aggressive patient behavior in medical facilities. Recording such incidents would allow for an accurate assessment of the scale of the phenomenon and perhaps persuade institutions and state authorities to take radical measures to reduce the risks posed by aggressive patient behavior in EDs. These measures would be complemented by the creation of an efficient information circulation system that would guarantee access to data on patients who engage in aggressive behavior.

Indirect aggressive behavior, especially actions carried out via the Internet (cyberbullying), should not be underestimated. Therefore, the police should be obliged to assist in **determining the IP address** of the computer from which a defamatory post was made on a website. Otherwise, it will not be possible to eliminate incidents of indirect aggressive behavior by patients.

Research has shown how dangerous patients who have consumed excessive amounts of alcohol are to medical staff in the EDs. Therefore, it would be advisable to establish a dedicated facility in the vicinity of the emergency room to care for people under the influence of alcohol. The creation of a **separate facility** would help to reduce the number of such patients waiting in the emergency room and thus shorten the waiting time for services. A helpful tool would be the obligation to introduce a **video surveillance system** in all EDs and a system for directly notifying the police, as well as equipping medical personnel with invisible cameras worn on their clothing, which would be activated in critical situations to record the incident. It would also be advisable for hospitals to have their own security services or hire a security company. This would ensure the safety of employees and the protection of property in the ED.

One of the key preventive measures from the point of view of medical staff safety, which does not require financial outlay, is the close **presence of colleagues when providing medical services in the emergency department**, especially in the evening and at night, when aggressive behavior on the part of patients is most prevalent. Another important aspect is raising the awareness of ED medical staff about the possibility of using the legal protection provided for public officials. In practice, it is possible to defend oneself against patients exhibiting aggressive behavior in the ED through criminal and civil law. However, these are overly formalized processes that require a series of procedures, time, determination, and significant activity on the part of the victim, both in the preparatory and trial proceedings.

The key issue is the need to **introduce courses** on dealing with patients exhibiting aggressive behavior into the curricula of medical universities. Training in dealing with agitated patients, interpersonal communication, coping with stress, self-defense, and applicable law is valuable. Such training should be offered to employees on a regular basis during their employment. Knowledge of the rules for dealing with aggressive patients will allow for an appropriate response and minimize the negative effects of such behavior. It will not reduce the number of acts of aggressive behavior by patients, but it will eliminate the feeling of helplessness among medical staff and significantly improve the effectiveness of defense against the threats resulting from such behavior.

The occupational risk assessment cards for medical personnel in direct contact with patients need to be refined so that they not only take into account aggressive behavior by patients but also contain information on the risk of its various forms occurring. Therefore, there is a need for research on aggressive behavior by patients toward doctors and nursing staff in other hospital wards. The rationale for proposing such research is based on the recommendations of respondents, primarily health and safety experts, who reported that aggressive patient behavior should be included in the occupational risk assessment cards of every employee representing the medical profession and having direct contact with patients. Despite comments from health and safety specialists regarding the expansion of occupational risk assessment documentation, such an assessment should be carried out again, taking into account the actual time spent by medical staff in direct verbal and physical contact with conscious patients. This would reveal the risk of exposure of doctors and nursing staff to aggressive behavior by patients and would allow occupational risk assessment cards to be adapted to the conditions prevailing at individual workstations.

The results of the study showed a lack of psychological support for employees who were subjected to aggressive behavior by patients. Therefore, it is worth implementing a **support strategy for victims** of aggressive behavior by patients. A lack of awareness and commitment on the part of the employer may be a barrier to reporting cases of aggressive patient behavior in the ED. In the context of aggressive behavior, it is essential **to educate patients** about their rights and how the ED works. Misunderstanding and ignorance of the rules often lead to conflict situations, resulting in aggressive behavior on the part of the patient and their family. It is particularly important to know about the right of priority to medical care for patients based on their health condition rather than the time of arrival at the ED. It would be helpful for medical facilities to cooperate with institutions working to protect the health of employees in order to conduct social campaigns to educate patients about the functioning of hospital EDs and the status of public officials enjoyed by medical staff in the ED.

Key recommendations for managing the risk associated with aggressive patients, which are important from the point of view of the safety of medical personnel in the ED, have been divided into recommendations implemented at the central (national) and organizational levels.

Recommendations for improving work safety in EDs at the state level:

- Introduction of legal provisions requiring employers to take measures to limit the escalation of aggressive behavior by patients in the ED and to protect medical staff from aggressive behavior by patients.
- Introduction of stricter penalties for aggressive behavior toward ED employees.
- Introduction of an obligation to record incidents of aggressive behavior by patients and creation of a central register of such behavior.
- Strictly assigning tasks to the police related to assisting in holding aggressively behaving patients accountable.
- Establishment of an ED dedicated to the care of persons under the influence of alcohol or other psychoactive substances or addicted to them.
- Introducing content and workshops into medical staff training programs that would comprehensively prepare medical staff to deal with patients exhibiting aggressive behavior, including training in the early recognition of dangerous situations, communication training aimed at de-escalating conflict, and self-defense training.

Recommendations to increase the safety of medical staff in EDs at the organizational level:

- Implementation of an occupational health and safety management system.
- Creation of internal procedures (instructions/guidelines) aimed at protecting medical staff from aggressive behavior on the part of patients.
- Updating occupational risk assessment cards due to the varying degrees of exposure of employees to aggressive behavior by patients, employed in the same positions, and updating remuneration in line with the risk incurred.
- Creating a culture of safety through education and raising awareness of safety among employees at various levels and patients.
- Develop simplified procedures for cooperation with the local police based on government legislation.
- Establishment of in-house security services or hiring of security companies.
- Establishment of a video surveillance system in hospital EDs and a system for directly notifying the police.
- Equipping medical staff with cameras to record incidents of aggressive patient behavior.

- Cooperation with institutions working to protect the health of employees in social campaigns aimed at educating patients about the functioning of hospital EDs and the status of public officials enjoyed by medical personnel during emergency operations.
- Developing and implementing a support strategy for victims of aggressive patient behavior. As part of the program, monitoring the condition of employees who are victims of aggressive patient behavior and supporting them in their further work in terms of physical, emotional, and mental health.

In counteracting patient aggression toward medical staff, it is important to have the efforts and cooperation of many social partners and to implement procedures and tools simultaneously at many levels: national, sectoral or regional, organizational, team, and finally, individual employees.

Comprehensive Strategies to Counteract Patient Violence in Emergency Departments

Research shows that implementing isolated procedures is less effective than a comprehensive approach focusing on the implementation of a variety of complementary tools at every stage of patient care in the ED. Such a strategic approach should focus on three stages:

- preventing acts of violence,
- actively responding when they occur, and
- mitigating the negative consequences of incidents that have already occurred.

This will help to create a culture of safety at work in the EDs and significantly improve the situation in the long term.

At the organizational level, it is important to involve employees from various departments, including ED medical staff, as well as management, security, and administrative personnel. Only joint action at various levels can bring about the desired results and improve safety.

When talking about the organizational level, the main focus is on appropriate integration in three aspects [Ramacciati et al. 2016]:

- improving the work environment,
- implementing procedures, and
- training employees.

The policy implemented must be transparent and known to employees. It is worth setting up an interdisciplinary team of employees to develop and implement the best solutions. This policy should regulate the expected behavior of employees and patients, the consequences of aggression/violence, reporting mechanisms and the protocol after a reported incident, as well as the support offered to employees after the incident. Examples of tools implemented in the first stage are shown in Table 6.2.

TABLE 6.2
Policy for Counteracting and De-Escalating Aggressive Patient Behavior – First Stage

Objective/ Target Group	Physical Environment	Intangible Work Environment
Patient-focused activities	Weapon and sharp instrument detection systems Electronic systems indicating the waiting time for intervention by the emergency department (ED) team (according to triads)	Information on legal and criminal proceedings in the event of aggression against ED staff Special "zero tolerance" procedures and signage in areas particularly vulnerable to aggression in the ED Occupational risk assessment and use of dedicated risk assessment tools such as the Aggressive Behavior Risk Assessment Tool (ABRAT), Violence Risk Assessment Tool (M55), Staring, Tone and volume of voice, Anxiety, Mumbling, and Pacing (STAMP), and Assessment, Behavioral Indicators, and Conversation (ABC)
Measures targeting employees	Appropriately designed waiting rooms in the ED, separating ED employees from the waiting room, providing good visibility of patients and easy access to the waiting room for employees Installation of a 24/7 camera surveillance system in key locations in the ED Installation of special buttons (panic buttons) or alarm systems for use by medical personnel in emergency situations	Promoting a culture of safety at work as part of creating a sustainable working environment (at the organizational level) for teams and employees Raising awareness of preventing and stopping violence in the ED (changing the mindset about patient aggression as part of the job) Involving employees, especially management, in the process of promoting a culture of safety in the workplace Encouraging medical staff to report incidents of violence and supporting the process of change in responding to acts of aggression Joint development of procedures within the organization in this area Simplifying incident reporting procedures Training of emergency room staff in counteracting, behaving during, and reporting incidents of violence in the emergency room, and dealing with the consequences of such incidents

Source: Own work based on the literature: Ghosh, et al., 2019, Lee et al. 2023, Meyer et al. 1997, Okundolor et al. 2021, Rabin et al. 2025, Tarczon, et al., 2025.

Preparing the Physical Work Environment

Among the good practices aimed at reducing patient aggression are **weapon detection systems**. The use of these systems in the EDs studied has led to the confiscation of over 10,000 dangerous tools, with visits during which these tools were confiscated accounting for only 3% of all visits to these EDs [Rabin et al. 2025]. Research conducted in the area of weapon or sharp instrument detection systems installed in hospitals have shown that medical staff also feel safer when such systems are installed (73% of respondents) [Meyer et al. 1997]. Such solutions increase the sense of security of ED staff, raise public awareness, and significantly hinder the use of sharp metal tools against medical staff. However, the cost of installing such detectors is quite significant, as it also involves the need to hire additional staff.

Another solution used to counteract aggression in the ED is to design the ED waiting room in such a way as to separate ED staff providing services to patients at a given moment from the waiting room. At the same time, it provides good visibility of the patients and easy access to the waiting room for staff so that they can intervene quickly at any time.

When it comes to waiting rooms, it is worth preparing special rooms for patients who pose a potential threat, whenever possible. These rooms are usually appropriately equipped and devoid of equipment that could be used to assault ED staff.

Another tool is electronic systems for indicating the waiting time for ED team intervention (according to the triad system), informing patients of the potential waiting time for their admission. This is one of the key ways to reduce aggression, as many incidents are caused by excessive waiting times for treatment.

Research shows that reducing waiting times or clearly informing patients about waiting times, which can be achieved through appropriate electronic systems or by a person registering patients and informing them about the possible time of treatment, has an impact on reducing cases of aggression in the waiting room [Rabin et al. 2025]. Similarly, informing patients about staff shortages or longer medical procedures being carried out at a given moment. In each of these cases, ongoing communication with the patient and their loved ones is crucial.

In turn, when looking at the support of medical staff, it seems necessary to install a system of surveillance cameras (24 hours a day, 7 days a week) in key locations in the ED. It would also be very helpful to install special buttons or alarm systems that medical staff can activate in an emergency.

Unfortunately, solutions related to the physical work environment are quite costly, but they do reduce the number of incidents and increase the sense of security among ED medical staff [Meyer et al. 1997].

Among the low-cost and relatively effective tools is a "zero tolerance" policy for aggression, especially when combined with a broader information campaign at the national level. However, at the organizational level, it is crucial to put up posters in waiting rooms and other emergency department areas to send a clear message that inappropriate behavior will not be tolerated and will be reported and punished accordingly. This is an important signal not only to potential aggressors but also to medical staff, that violent behavior is not the norm and will not be tolerated, and that such incidents must be reported.

Intangible Work Environment

Early prediction and mitigation of workplace violence can improve the safety of ED staff and the quality of care. To prevent workplace violence, ED staff must recognize and continuously monitor risk factors for workplace violence from admission to the ED to discharge from the department. Research indicates that workplace violence can be predicted based on previous data on ED visits and stays documented in electronic medical records [Lee et al. 2023]. Various tools can be used for this purpose, such as Assessment, Behavioral Indicators, and Conversation (ABC), Aggressive Behavior Risk Assessment Tool (ABRAT), or Violence Risk Assessment Tool (M55). Currently, the most commonly recommended tool is Staring, Tone and volume of voice, Anxiety, Mumbling, and Pacing (STAMP), which is effective in assessing the potential risk of aggression from a patient at an early stage. The acronym stands for observed behavioral signs in five areas: Staring, Tone and volume of voice, Anxiety, Mumbling, and Pacing. This model has been expanded by adding four more domains: Emotions, Disease process, Assertiveness/non-assertiveness, and Resources as potential predictors of aggressive behavior (STAMP/EDAR model). It appears that the STAMP/EDAR model will be easy to use by nurses working in intensive care units without prior knowledge of the patient's history [Ghosh et al. 2019].

Strategies for counteracting patient aggression that are implemented in the ED may, in addition to the tools discussed above, include screening checklists used to identify signs of aggressive behavior at an early stage of escalation. This simple tool can be very useful, so a sample checklist is provided below (Table 6.3).

Studies have shown that the use of risk assessment checklists and prevention protocols has significantly reduced the number of incidents of aggressive patient behavior. Therefore, it is important that ED medical staff are familiar with the use of these tools as a method of assessing patients in EDs and reducing the risk of violence at an early stage. It is suggested that all patients or their companions at high risk of violence be assessed using a checklist upon admission and that the result be entered into their medical records. In addition, other members of the healthcare team could be informed about patients at high risk of violence [Sharifi et al. 2019].

Medical staff should be trained to recognize early signs of patient aggression in order to respond appropriately and prevent conflict escalation by identifying factors affecting patients that may lead to aggression in the long term [Rabin et al. 2025].

An important part of the strategy for counteracting patient violence in the ED is promoting a culture of safety at work as part of creating a balanced working environment. To this end, measures implemented at the organizational level may include

TABLE 6.3
Sample Checklist for Assessing the Risk of Patient Aggression

No.	Risk Factor	Points
1.	Previous acts of aggression by the patient	1
2.	Suicidal or aggressive thoughts	1
3.	Aggressive intentions	1
4.	Abuse or influence of psychoactive substances or alcohol	1
5.	Traumatic experiences (PTSD, exposure to trauma, intellectual disability)	1
6.	Lack of insight/judgment	1
7.	Emotional instability/unpredictable behavior or mood swings	1
8.	Employee intuition/other factors not listed	1-2

Source: Based on Okundolor et al. [2021].

raising awareness among all stakeholders about violence in the workplace. It is crucial that management is strongly committed to promoting a culture of safety in the workplace, as it is their actions and commitment that lead to significant changes in how workplace violence (WPV) is responded to. It is also important to inform patients and their relatives about the unacceptability of aggressive behavior and the potential penalties for such behavior. To this end, so-called "zero tolerance" procedures and signage are being introduced in areas particularly vulnerable to violence. Such a procedure should be developed in a given unit by employees from various levels of management.[10]

Elements of creating a culture of safety in this area will also include activities focused on changing the mentality of medical staff in terms of perceiving violence in the ED as an integral part of their work. When employees stop identifying acts of aggression in the ED with their work and start perceiving such incidents as something that should never happen, this will accelerate the elimination of such incidents and their condemnation and reporting. However, it is important to support employees in this process, e.g., by setting up special teams to respond appropriately to emerging aggression or to subsequently investigate, enforce penalties, and support employees.

A culture of safety should also be built through appropriate training of employees in the use of the aforementioned standardized tools and procedures for early recognition of potential threats, as well as procedures and systems for reporting cases. Training employees at every stage of their professional career (preparation in medical school, before starting work, or during work) is a key element in supporting their safety in their daily work.

At this level, it is recommended to establish an interdisciplinary team to respond to emergency situations, which will intervene and support medical staff during incidents. Such a team should include people from various professions within the facility, e.g., medical and administrative staff, lawyers, security personnel, and management. In an emergency situation, the appointed team will communicate patients' obligations to them and discipline and enforce compliance with the law. Special training and experience in dealing with acts of violence will enable the team to take a professional approach and act effectively, as well as provide real assistance to ED staff. It is also necessary to develop effective communication channels and codes that will be used in an emergency to quickly notify security personnel and the response team.

Moreover, it is recommended to build internal responsibility among medical staff for reporting every act of aggression against employees and for taking action to intervene in potentially dangerous situations. Unfortunately, studies conducted in some countries indicate that medical staff are reluctant to report acts of violence [Martinez 2016; Tung et al. 2021, Tarczoń et al. 2025]. The emerging proactive behavior of ED staff, such as requests for support, appropriate alarm systems, or staff ideas for de-escalating patient aggression in the ED, indicate a growing awareness among staff and a dissatisfaction with the current situation and a need for change. This is a good signal to work together within the facility to develop procedures and tools tailored to the needs of employees and the specific nature of a particular ED.

TABLE 6.4
De-Escalation Policy and Actions to Be Taken when Aggressive Behavior Occurs in the Emergency Department – Second Stage

Objective/ Target Group	Physical Work Environment	Intangible Work Environment
Patient-focused actions	Special rooms prepared for the separation of patients posing a threat in the ED	Additional staff to accompany patients and inform them about the triage procedure and waiting times during periods of high patient volume Refusal to admit or informing patients with a history of aggressive behavior toward ED staff about conditional admission and applying the highest precautions toward them – e.g., special rooms and security presence Enforcement of penalties for persons who commit violence
Employee-focused measures	Support for security personnel Employee training Training materials dedicated to unit employees (developed on the basis of incidents and possible response scenarios) Periodic online training reminding employees of basic safety rules, tests, etc.	Establishment of an interdisciplinary team to respond to emergency situations Development of protocols for employees to help communicate the need to enforce the law in the event of incidents Development of fast and effective communication channels between employees and teams activated during times of high pressure at work Creation of a special communication code to call security personnel in case of danger Building personal responsibility for reporting any act of violence or noticing a potentially dangerous situation Training dedicated to medical staff on self-defense, de-escalation of conflict situations, non-violent communication, ways of dealing with the consequences of patient aggression, etc. Dialogue with employees on the development of procedures and tools dedicated to a given unit. Creating codes of ethics that include aspects of mutual respect and building a culture of safety at work

Source: Own work based on the literature: Ghosh, et al., 2019, Lee et al. 2023, Meyer et al. 1997, Okundolor et al. 2021, Rabin et al. 2025, Tarczoń, et al., 2025.

One of the key tools at this stage is also intensive, active training in communication, conflict management, and tension relief, which should be repeated systematically. There is also a need for workshops and simulations on managing patients prone to violence, responding appropriately to such behavior, and self-defense training. The latter is rarely conducted, but it gives employees a sense of security and practical skills for dealing with situations of physical danger. Research confirms that training employees gives them a sense of confidence and helps them to better cope with patient aggression [Rabin et al. 2025]. Particular emphasis is placed on training in the simulation of medical team work during an act of violence, with a precise division of roles within the medical team in the event of such an incident.

TABLE 6.5
Policy for Mitigating the Negative Consequences of Incidents of Aggressive Patient Behavior in the Emergency Department – Third Stage

Objective/Target Group	Physical Environment	Intangible Work Environment
Activities focused on employees and improving the safety culture	Employee suggestions and proposals regarding the physical work environment based on an analysis of incidents that occurred during the year	Entry of data on the patient or person who committed the act of aggression – a register to avoid similar incidents in the future Interviewing employees who have experienced aggression Providing immediate necessary support to affected employees (e.g., medical and psychological) Possibility of longer-term therapy or counseling for victims Observation of the behavior and needs of victims for several weeks after the incident Case analysis by the team and conclusions for the future Preparation and implementation of scenarios for conversations with employees or teams experiencing aggression/violence from patients and their relatives – training of the team to conduct interviews Improvement of procedures, including incident reporting

Source: Own work based on the literature: Ghosh, et al., 2019, Lee et al. 2023, Meyer et al. 1997, Okundolor et al. 2021, Rabin et al. 2025, Tarczoń, et al., 2025.

Despite improvements in procedures and efforts to measure their effectiveness, healthcare workers perceive the actions taken by employers after reporting an incident of aggression as insufficient [Brophy et al. 2018]. Therefore, employers should ensure that employees feel supported by the organization. Employees' confidence that their employer is striving to create a supportive and safe working environment and is making efforts to improve the situation can significantly influence changes in employee behavior and their sense of purpose in reporting incidents. The key is how to proceed after an incident, e.g., interviewing employees, filling out reports, offering support to injured employees, or punishing perpetrators. Support should include not only post-incident interviews but also medical and psychological support for victims. Rapid and comprehensive support (including emotional support) will help to mitigate the long-term effects of violence, reduce stress, and aid in the healing process. There is also usually a lack of formal procedures for reporting and monitoring cases of patient violence. Unfortunately, most often, EDs lack systematic and structured training in de-escalating workplace violence, as well as a formal structured interview after an incident of patient aggression [Rabin et al. 2025].

Observing the upward trend in this phenomenon, it is worth considering appropriate training in de-escalating violence and dealing with acts of violence at the education stage, both in medical studies and later when starting work in the ED. Most medical staff, whether doctors, nursing staff, or paramedics, have direct contact with patients, so this seems natural. Similarly, training for management staff in medical facilities should address issues related to creating a balanced work environment and developing competencies related to dealing with patient aggression. Among the most effective forms of improving such skills is a **blended approach**, starting with remote training combined with simulations in which employees respond to simulated acts of violence. Only practical simulations teach quick and appropriate responses and allow employees to experience the stress and emotions that arise during a violent incident [Rabin et al. 2025]. What is more, training conducted for employees from different departments simultaneously helps them understand other perspectives and supports subsequent cooperation during such incidents. Self-defense training for employees, which builds personal agency and a sense of security for medical staff, seems particularly important, although it is still not popular among employers. Research shows that only about 16% of employees have training in de-escalating aggression and only 10% in self-defense [Rabin et al. 2025].

SUMMARY

A sustainable work environment and safe employee behavior in the workplace do not develop automatically. They must be developed by taking measures to eliminate threats and introducing the necessary changes, which must be improved and maintained. It seems important not only to mitigate the negative effects of insecurity in hospital EDs but also to shape that security in real terms through actions taken at the state, local, and organizational levels. Strategies that are effective in the long term and lead to the elimination of aggression in the workplace should integrate a number of tools at different levels, involve key stakeholders who continuously improve existing processes through dialogue, and measure and evaluate effectiveness. Employee participation is crucial in this context. The main argument in favor of implementing employee participation is concern for the common good, which manifests itself in an increased role for employees, contributing to the development of the organization. Ideally, these activities, which are part of, should be treated as a priority in the culture of occupational safety, which will translate into the well-being of employees as well as patients and their families.

The problem of lack of support for medical staff in the ED and "permission" for aggressive behavior by patients is part of a broader discourse on creating a culture of workplace safety or the low social status of important professions in some countries [Andrałojć 2023].

The results of the study suggest the need for further research and analysis to enable changes aimed at reducing the escalation of aggressive behavior by patients toward medical staff providing medical services in EDs. In an attempt to identify directions for future research on the phenomenon of aggressive patient behavior, there is a need for more advanced research involving medical staff from all EDs in different countries. It is proposed that research be conducted on the exposure of other professional groups (e.g., registrars, ward assistants, paramedics, and medical caregivers) employed in the ED to aggressive behavior by patients.

NOTES

1 Regulation of the Minister of Health in Poland of April 22, 2005.
2 Act of October 30, 2002.
3 https://www.england.nhs.uk/supporting-our-nhs-people/health-and-wellbeing-programmes/violence-prevention-and-safety/. Accessed on October 30, 2024.
4 https://www.england.nhs.uk/wp-content/uploads/2022/06/B0989-NHS-violence-prevention-and-reduction-standard-guidance-notes.pdf. Accessed on May 10, 2025.
5 https://www.cdc.gov/niosh/index.html. Accessed on May 10, 2025.
6 https://www.aha.org/system/files/media/file/2019/08/mha-developing-healthcare-safety-violence-prevention-programs-in-hospitals.pdf. Accessed on 12.04.2025.
7 *The Royal Melbourne Hospital* (https://www.thermh.org.au/your-care/hospital-care/what-you-need-to-know-up-front/aggressive-violent-behaviour 17.11.2023).
8 *Department of Health Victoria, Addressing violence in Victorian healthcare settings* (https://www.health.vic.gov.au/worker-health-wellbeing/occupational-violence-and-aggression-health-facility-design 02.01.2026).
9 *Emergency Care in Victoria* (https://www.health.vic.gov.au/patient-care/emergency-care 11/27/2023).
10 An example of a procedure and information campaign conducted by the NHS. https://www.essingtonmedicalcentre.nhs.uk/policies/zero-tolerance-policy/.

Conclusion

Violence and abuse toward medical staff are one of the many factors that can have a devastating and lasting impact on their health and well-being. Many countries still lack good practices and systemic solutions aimed at reducing the number of incidents involving aggressive behavior by patients in the healthcare sector. It is therefore necessary to take measures to limit the escalation of aggressive behavior by patients and to build a safe working environment for doctors, nursing staff, and paramedics, not only in hospital emergency departments (EDs).

Ensuring an adequate level of safety and health protection for employees brings many benefits to the company, although, it is often perceived as generating additional, unnecessary costs resulting from expenditure on collective and individual employee protection measures. Meanwhile, investments in occupational health and safety reduce the costs of running a business by reducing sick leave resulting from occupational diseases and accidents at work [Amoako-Asiedu et al., 2023], which in turn allows the company to maintain continuity of operations.

Coordinated action at the government level in individual countries, at the organizational level, and at the team and individual levels seems necessary. A broad information campaign to raise awareness and condemn patient aggression toward medical staff is also crucial. This problem is not well recognized by the public, which may exacerbate the current situation. Public awareness in this regard and public support for medical workers can help in the fight against such acts of violence. Therefore, it should be shaped in order to build public support for medical staff and reinforce the sense of injustice by establishing social norms in which such behavior is unacceptable. Such social sensitization can help shape appropriate behaviors and attitudes of patients in the emergency room.

However, regardless of whether such measures are taken in a given country, organizations can try to take important steps on their own. It is important, however, that measures introducing change are properly planned and controlled. It is recommended that the measures taken by employers be consistent and comprehensively address the promotion and protection of employee health, using, for example, the *Total Worker Health* approach. This approach has been proven effective in studies due to its direct impact on employee health, their engagement and active participation in such programs, and cost reduction [Sorensen et al. 2016]. *Total Worker Health* encompasses all aspects of work within integrated interventions that address safety, health promotion, and employee well-being [NIOSH 2023[1]]. To optimize the effectiveness of interventions, greater emphasis should be placed on the development and implementation of safety initiatives in a broader organizational context, involving occupational safety management and the development of a safety culture within the organization. Therefore, it is proposed to introduce interventions at the organizational, team, and individual levels. Such a comprehensive approach allows increasing the benefits of the implemented interventions; it also requires monitoring their

DOI: 10.1201/9781003716631-8

effectiveness, which is time-consuming and costly, but in the long term will allow this area to be managed.

In addition, to achieve the best results, the authors propose introducing **interventions at three stages** [Sinclair et al. 2010]:

- **First-stage interventions**, which focus on eliminating the problem and preventing or reducing the occurrence of aggressive behavior in patients.
- **Second-stage interventions**, which support employees in dealing with the existing problem of aggressive patient behavior, promote health-promoting practices, and help increase the individual resources of medical staff (skills to deal with incidents of aggressive patient behavior).
- **Third-stage interventions**, leading to the removal and elimination of the effects of aggressive patient behavior, psychological and medical assistance to employees necessary to restore their health.

First-stage measures may include, for example, simplifying incident reporting processes and procedures, implementing new regulations for the protection of medical staff, and creating social support systems in hospitals (cooperation between professional groups – cooperation with security and the police). These, in turn, may include proven solutions such as a zero-tolerance policy for violence, the preparation of isolation rooms, monitoring of injured employees for up to three months after the incident, support groups in the workplace where experiences and ways of dealing with cases of aggression can be shared, committees for counteracting incidents of aggressive behavior by patients, dialogue with employees in this area, work analysis taking into account new risk factors in the workplace (e.g., risk registers), work reorganization, close cooperation with the police, and letters and messages addressed to patients who behave aggressively and recording them in patient records.

Second-stage measures aimed at increasing the competence of hospital staff and their ability to deal with incidents of aggression may include, for example, self-defense training, training related to the process and methods of reporting incidents, mediation training, and raising staff awareness of their rights.

Third-stage measures, in turn, may include trauma counseling, additional days of leave after such an incident, additional medical and pension insurance for employees, psychological support, higher remuneration or a financial allowance to cover the damage incurred, care for the physical work environment, and others.

Actions carried out at all three stages require appropriate planning and implementation of programs, but such a strategic and comprehensive approach can bring measurable results for all stakeholders: employers, employees, and patients. Such a comprehensive approach can realistically increase the safety of medical workers, as it is linked to the development of a culture of safety, i.e., improving the processes and procedures for reporting such incidents, organizational support for employees in potentially dangerous situations, but also shaping responsible employee behavior related to, for example, improving competencies in this area, eliminating emerging risks in this area, and reporting incidents.

There is little research on the effectiveness of de-escalating aggressive patient behavior, and there are no standards of practice. Medical staff are given freedom of action, which unfortunately does not bring the expected results or serve the interests of individual stakeholder groups.

In the context of a shortage of medical staff, more and more attention is being paid to the well-being of employees and the creation of a supportive work environment. The medical sector is one of the most dangerous work environments, where, in addition to classic hazards, there are psychosocial risks, including aggressive patient behavior. Aggressive patient behavior strongly interferes with the well-being of medical staff, negatively affecting their well-being and health, commitment to work, and even work performance, as well as disrupting the work of medical facilities.

According to research, the most common causes of aggressive patient behavior in EDs are patient intoxication, psychoactive substance use, nicotine cravings, but also a lack of understanding of the situation in which patients find themselves and negligence in treatment and care. ED medical staff most often experience aggressive behavior from patients in verbal form, less often in physical form, although the latter has the most severe consequences. Psychological forms of aggressive behavior in the form of intimidation and threats directed at medical staff are also becoming more common. In addition to these direct forms of aggressive behavior, there are also indirect forms, such as offensive text messages and emails, or false posts on social media.

Unfortunately, the results of the research show that incidents of aggressive behavior by patients toward medical staff in EDs occur very often and their consequences can be serious and should be considered on many levels:

- individual (employee dissatisfaction with work, loss of dignity and professional pride, long-term stress, and general health problems);
- organizational (financial losses for the medical facility, low morale and dissatisfaction among employees, disruption of work, high risk of medical errors and reduced quality of services, as well as decreased work efficiency); and even
- social (effects on the family of the affected employee, other patients, and medical staff).

Given the scale and frequency of such incidents, this threat should be included in the occupational risks of medical professions, and it seems necessary to monitor such events. Research indicates that there is a lack of genuine interest on the part of employers in reporting such incidents and a lack of necessary support in the form of incentives and procedures for dealing with aggressive patient behavior in the emergency. This is the main reason, alongside the lack of legal regulations, for the sporadic reporting of cases of aggressive patient behavior by employees. Based on the results of the study, a number of practical recommendations have been proposed to help reduce this risk at both the state and organizational levels. Analyzing the organizational possibilities, the need for intervention on three stages was identified: to prevent such incidents, to support employees during incidents, and to mitigate the effects of incidents that have already occurred.

Creating a safe and supportive work environment is only one of the key aspects of ensuring smooth operation in healthcare. The problem of lack of support for medical staff in EDs and "permission" for aggressive behavior by patients is part of a broader discourse on creating a culture of safety at work or the low social status of important socially professions in some countries. It is necessary to be aware that performing a demanding and important medical profession in the face of low social status and a high sense of threat creates a negative image not only of the facility but also of the medical professions. This can have a huge impact on the outflow of medical workers, which, in the context of an aging population and a growing shortage of workers in this sector, can be a serious social problem.

The authors of the book hope that the research results will increase public awareness in this area, and the recommendations contained therein will prove helpful in managing healthcare workers and creating a supportive work environment for them that will help create the well-being of both employees and patients.

NOTE

1 NIOSH 2023, *Total Worker Health Program*, Centers for Disease Control and Prevention (https://www.cdc.gov/niosh/twh/default.html 23.03.2024).

Appendix
Survey Questionnaire

Part I consists of questions about incidents of aggressive behavior by patients and accompanying persons, their type, and frequency.

1. Have you ever had contact with an aggressively behaving patient?
   - Yes
   - No
2. If you have had contact, what type of aggression was it?
   - Physical
   - Psychological
   - Verbal
   - Difficult to say
3. How often did you have contact with an aggressive patient?
   - During every shift
   - During most shifts
   - Approximately every other shift
   - Once in a while
   - Never
4. What forms of physical aggression have you encountered and how often did they occur? (You can give one answer per line.)

Frequency Form of Aggression	During Every Shift	During Most Shifts	Roughly Every Other Shift	Once in a While	Never
Shaking					
Hitting/attempted assault					
Kicking					
Biting					
Pushing/shoving					
Waving arms					
Hair pulling					

5. What other forms of physical aggression have you encountered?

   ………………………………………………………………………………………

6. What forms of psychological aggression have you encountered? (You can give one answer per line.)

Frequency Form of Aggression	During Every Shift	During Most Shifts	Roughly Every Other Shift	Once in a While	Never
Blackmail					
Threat					
Intimidation					
Ridicule					
Humiliation					
False public accusation, slander					
Persistent harassment (reporting, written complaint)					

7. What other forms of psychological aggression have you encountered?

   ……………………………………………………………………………………

8. What forms of verbal aggression have you encountered? (You can give one answer per line.)

Frequency Form of Aggression	During Every Shift	During Most Shifts	Roughly Every Other Shift	Once in a While	Never
Shouting					
Verbal skirmishes					
Insults					
Swearing					
Defamation					
Arguments					
Sweers					

9. What other forms of verbal aggression have you encountered?

   ……………………………………………………………………………………

10. Have you ever experienced indirect aggressive behavior, such as?
    - Offensive text messages and emails
    - Posts on social media
    - Publication of humiliating, false information on the Internet
    - Posting photos or videos ridiculing the victim

CAUSES AND CONSEQUENCES

Part II consists of questions about the causes of aggressive behavior in patients and those accompanying them, issues resulting from such behavior, and how to deal with patients who behave aggressively.

11. In your opinion, what are the causes of aggressive behavior in patients?
    a. Alcohol
    b. Psychoactive substances (intoxicants)

c. Pain and suffering due to illness
d. Difficult life situation
e. Lack of security
f. Negative assessment of medical staff actions
g. Other (what?)

For each statement, please indicate to what extent you agree with it on a scale of 1 to 5 (1 – "I completely disagree"; 2 – "disagree"; 3 – "no opinion"; 4 – "agree"; 5 – "strongly agree").

12. I believe that aggressive patient behavior is an inherent part of working in the emergency room
    I completely disagree 1 2 3 4 5 I completely agree
13. The patient behaves aggressively due to limited access to medical services
    I completely disagree 1 2 3 4 5 I completely agree
14. The patient is acting aggressively because they have been waiting too long for medical advice
    I completely disagree 1 2 3 4 5 I completely agree
15. The patient behaves aggressively because of dissatisfaction with the range of medical services offered
    I completely disagree 1 2 3 4 5 I completely agree
16. The patient often behaves aggressively when medical procedures are being performed
    I completely disagree 1 2 3 4 5 I completely agree
17. The patient behaves aggressively because of the refusal to issue a certificate of incapacity for work, a medical certificate, a prescription, a referral for tests, etc.
    I completely disagree 1 2 3 4 5 I completely agree
18. The patient behaves aggressively due to the need for inpatient treatment – hospitalization
    I completely disagree 1 2 3 4 5 I completely agree
19. In the ward where I work, employees are blamed for patients' aggressive behavior
    I completely disagree 1 2 3 4 5 I completely agree
20. Aggressive behavior by patients during their stay in the emergency room is justified
    I completely disagree 1 2 3 4 5 I completely agree
21. No incidents of aggressive patient behavior during their stay in the ER are recorded in any documentation
    I completely disagree 1 2 3 4 5 I completely agree
22. The employer is interested in information about aggressive behavior by patients in the emergency room
    I completely disagree 1 2 3 4 5 I completely agree
23. Are there any specific situations that cause an increase in aggressive behavior by patients toward medical staff?
    - Yes (what kind?)
    - No
    - Difficult to say

24. What are your reactions to aggressive patient behavior?
    - Ignoring the behavior
    - Discussing it with the patient
    - Asking the patient to change their behavior
    - Use of sedative treatment
    - Forced immobilization
    - Seeking help from the police
    - Asking colleagues for help
    - Other (what?)
25. At what time of day do patients most often behave aggressively?
    a. In the morning
    b. In the evening
    c. During the night
    d. During the day
    e. In the middle of the day
    f. Hard to say
    g. Other
26. Could the patient's aggressive behavior be a result of the staff's actions?
    a. Yes
    b. No
    c. Difficult to say
27. If so, what actions of ER staff could cause aggressive behavior in patients?
    a. Lack of time to explain the situation and procedures
    b. Lack of understanding of the patient's situation
    c. Downplaying the patient's symptoms
    d. Negligence in treatment and care
    e. Provocative behavior on the part of medical staff
    f. Other (what?)
28. Who, based on gender, is most often the victim of aggressive patient behavior?
    a. Women
    b. Men
    c. Difficult to say
29. Are patients more likely to behave aggressively in the presence of accompanying persons?
    a. Yes
    b. No
    c. Hard to say
30. If so, the presence of which accompanying persons increases the patient's aggressive behavior?
    a. Family members
    b. Partners
    c. Security services (police or security guards)
    d. Other patients
    e. Others (which ones?)
31. What emotions do you experience when dealing with an aggressive patient?
    a. Anger
    b. Anxiety and fear

c. Helplessness
d. Uncertainty
e. Rage
f. Sadness
g. Regret
h. Breakdown
i. Agitation
j. Other (what?)

32. What moods accompany you after contact with a patient who behaves aggressively?
    a. Long-term stress
    b. Loss of dignity and professional pride
    c. Low self-esteem
    d. Depressive states
    e. General health problems
    f. Job dissatisfaction
    g. Other (what?)..
33. Apart from patients, who else would you identify as perpetrators of aggressive behavior in your workplace?
    a. The patient's family
    b. The patient's friends
    c. Colleagues
    d. Superiors
    e. Subordinates
    f. Other people (who?)..
34. What methods do you use to relax after contact with an aggressive patient?
    a. Watching television
    b. Listening to music
    c. Being around animals
    d. Playing sports
    e. Meeting friends
    f. Calling a friend
    g. Household chores
    h. Helping others
    i. Other (what?)

ORGANIZATION AND MANAGEMENT

Part III consists of questions about the impact of aggressive patient behavior on the organization's activities and opinions on the need for training in dealing with aggression toward staff.

35. Is aggressive patient behavior included as a risk factor in your occupational risk assessment forms?
    - Yes
    - No
    - I don't know

36. In your opinion, how well is the emergency department prepared to prevent aggressive patient behavior?
    a. Very poor
    b. Poor
    c. Average
    d. Good
    e. Very good
37. Does the emergency room have effective procedures for responding to aggressive patient behavior?
    a. Yes
    b. No
    c. I don't know
38. Does the emergency department have a record of incidents involving aggressive patient behavior?
    a. Yes
    b. No
    c. I don't know
39. Does the emergency department analyze the causes of aggressive patient behavior?
    a. Yes
    b. No
    c. Hard to say
40. Does the hospital provide psychological assistance to victims of aggressive behavior by patients?
    a. Yes
    b. No
41. In your opinion, what are the organizational consequences of aggressive behavior by patients toward medical staff?
    a. Lower quality of services provided
    b. Decreased work efficiency
    c. Disruption of work
    d. Risk of medical errors
    e. Staff absenteeism
    f. Costs (renovation or purchase of new equipment)
    g. Other (what?)
42. Do you consider it necessary to include content on dealing with aggressive patients in the training of medical personnel (doctors, nurses, and paramedics)?
    a. Yes
    b. No
    c. Difficult to say

43. What kind of training would be particularly useful in order to increase the safety of employees when dealing with aggressive patients? (You can give one answer per line.)

Training	**Useful**	**Quite Useful**	**Not Useful**
Interpersonal communication			
Applicable law			
Self-defense techniques			
Dealing with agitated patients			

44. What other training would be particularly useful in order to increase employee safety when dealing with aggressive patients?

    ....

45. Do you benefit from the legal protection provided to public officials?
    a. Yes
    b. No
46. In your opinion, should entities be established that are required to inform patients about the legal protection available to medical personnel?
    a. Yes
    b. No
    c. Hard to say
47. In your opinion, is there a need for radical measures to limit the escalation of aggressive behavior by patients?
    a. Yes
    b. No
    c. Hard to say

PERSONAL DETAILS

1. Please indicate your gender
   a. Female
   b. Male
2. Occupation
   a. Doctor
   b. Nurse
   c. Paramedic
3. Please specify your length of service in the profession

   ....

4. Place of residence
   a. Village
   b. Town with up to 20,000 inhabitants
   c. City with a population of 20,000 to 100,000
   d. City with a population of 100,000 to 200,000
   e. City with a population of 200,000 to 500,000
   f. City with a population of over 500,000
   g. Other
5. Type of residence
   - Alone
   - With family
   - Other
6. Number of people living in your household

   ……………………………………………………………………………

7. Size of the organization where you work
   - Up to 500 employees
   - 500 to 750 employees
   - 750 to 1,000 employees
   - Over 1,000 employees
8. Form of employment
   - Permanent employment contract
   - Fixed-term employment contract
   - Substitute employment contract
   - Probationary period contract
   - Contract agreement
   - Contract of mandate
   - Other……………………………………………………………………
9. Place of work
   - City with up to 20,000 inhabitants
   - City with 20,000–100,000 inhabitants
   - City with 100,000–200,000 inhabitants
   - City with 200,000–500,000 inhabitants
   - City with a population of over 500,000
10. Please specify the province in which the medical facility is located

    ………………………………………………………………………………

Bibliography

Ackerman R.J., Pickering S.E., 2002, *Before It's Too Late: Violence and Control in the Family*, Gdańsk Psychological Publishing House, Gdańsk.

Adamska K., 2016, *Silence and Breaking Silence in an Organization*, Psychological Annals, vol. 19: 75–88.

Aladwan M., Dalahmeh E., 2022, *Prevention of Workplace Violence in ED Nursing Using the Implementation of an Educational Program and a New Reporting Tool*, Clinical Journal of Nursing Care and Practice, vol. 6: 001–008.

ALBashtawy M., Aljezawi M., 2016, *Emergency Nurses' Perspective of Workplace Violence in Jordanian Hospitals: A National Survey*, International Emergency Nursing, vol. 24: 61–65. doi: 10.1016/j.ienj.2015.06.005.

Alharthy N., Mutairi M.A., Alsahli A. et al., 2017, *Workplace Violence among Emergency Medical Services Workers in Riyadh, Saudi Arabia*, Journal of Hospital Administration, vol. 6, no. 3: 26–32.

Aljohani B., Burkholder J., Tran Q.K., 2021, *Workplace Violence in the Emergency Department. A Systematic Review and Meta-Analysis*, Public Health, vol. 196: 186–197.

Anderson C., Bushman B., 2002, *Human Aggression*, Annual Review of Psychology, vol. 53: 27–51.

Amoako-Asiedu E, Appiagyei J., Amfo-Otu R., Parku K., Obuobisa-Darko T.,2023, Occupational Health and Safety, Cost Reduction in Accident and Employee Task Performance: Perspectives of Selected Service Organizations, Journal of Public Health, vol. 33: 2011–2022. doi: 10.1007/s10389-023-02150-6.

Andrałojć M., 2023, *The Social Value of Work in the Light of Research on the Nature and Significance of Professional Work. Scientific Journals of the University of Upper Silesia (or a Similar Name, Potentially Zeszyty Naukowe Akademii Górnośląskiej in Polish)*, 5–14.

Armstrong M., 2007, *Human Resource Management*, 4th edition, Wolters Kluwer business, Krakow.

Armstrong M., Taylor S., 2016, *Human Resource Management*, 6th edition, Wolters Kluwer, Warsaw.

Bandura A., Walters R., 1965, After: Kumięga P., 2014, Theories of Aggression [in:] A. Steciwko, J. Baranski, A. Mastalerz-Migas (eds.), *Aggression in Healthcare. Introductory Issues*, Elsevier Urban &Partner, Wrocław.

Bartczak M., Bartczak M., 2010, *Stress Intensity and Ways of Coping with Stress among Paramedics and Emergency Medical Services Students*, Scientific Yearbook of the Kujawsko-Pomorskie Higher School in Bydgoszcz. Transdisciplinary Studies on Culture (and) Education No. 5, 7–17.

Bazerman M., Neale M., 1997, *Negotiating Rationally*, Polish Psychological Association, "Libra," Publishing House, Olsztyn.

Becker-Pestka D., Kołodziej J., Pujer K., 2017, *Personal and Professional Development. Selected Issues in Theory and Practice*, EXANTE Publishing House, Wrocław.

Beltrami E.M., Kozak A., Williams I.T. et al., 2003, *Transmission of HIV and Hepatitis C Virus from a Nursing Home Patient to a Health Care Worker*, American Journal of Infection Control, vol. 31, no. 3: 68–175.

Berkowitz L., 1962, after: Kowalczuk K., Jankowiak B., Krajewska-Kułak E. et al., 2011, *Theories of Aggression-Truth or Myths?* Nursing Problems, no. 19(3): 396–400.

Blanchard K., 2013, *Higher-Level Leadership*, PWN Scientific Publishing House, Warsaw.

Braun V., Clarke V., 2006, *Using Thematic Analysis in Psychology*, Qualitative Research in Psychology, vol. 3, no. 2: 77–101.

Brophy J.T., Keith M.M., Hurley, 2018, *Assaulted and Unheard: Violence against Healthcare Staff*, New Solutions, vol. 27: 581–606.

Brown K., Herbert M., 1999, *Prevention of Domestic Violence*, State Agency for Solving Alcohol-Related Problems, Warsaw.

Buchelt-Nawara B., 2006, *HRM in Medical Service Units* [in:] H. Król, A. Ludwiczyński (eds.), *Human Resource Management. Creating Human Capital in Organizations*, Wydawnictwo Naukowe PWN, Warsaw.

Burak A., Antoniewicz J., Malinowska E. et al., 2021, *Documenting Aggression by Medical Staff in Hospital Emergency Departments*, Nursing in the 21st Century, vol. 20, no. 2: 105–116.

Buss A.H., 1961, after: Kumięga P., 2014, *Theories of Aggression* [in:] A. Steciwko, J. Barański, A. Mastalerz-Migas (eds.), *Aggression in Healthcare. Introductory Issues*, Elsevier Urban & Partner, Wroclaw; Lickiewicz J., 2017, *The Aggressive Patient. Methods and Techniques of Management*, PZWL Medical Publishing House, Warsaw.

Cewińska J., 2016, *Fear in the Workplace and Its Negative Consequences*, Scientific Papers of the University of Economics in Wrocław, vol. 240, 49–57.

Cooper M.D., 2000, *Towards a Model of Safety Culture*, Safety Science, vol. 36, no. 2: 11–136.

Costa D., Mercieca-Bebber R., Tesson S. et al., 2019, *Patient, Client, Consumer, Survivor or Other Alternatives? A Scoping Review of Preferred Terms for Labelling Individuals Who Access Healthcare across Settings*, BMJ Open, (https://bmjopen.bmj.com/content/bmjopen/9/3/e025166.full.pdf 30.12.2025).

Cox T., Griffiths A., Rial-Gonzalez E., 2000, *Research on Work-Related Stress*, EU-OSHA, Luxembourg.

Czakon W., 2022, *The Use of Case Studies in Management Science Research* [in:] Czakon W. (ed.), *Fundamentals of Research Methodology in Management Science*, Wydawnictwo Nieoczywiste, Warsaw.

Czemierowska-Koruba E., 2015, *Aggression and Violence at School, or What We Should Know to Act Effectively*, 2nd edition, Center for Education Development Edition, Warsaw.

Czerw A., 2017, *Psychological Model of Well-Being at Work. The Value and Meaning of Work*, Wydawnictwo Naukowe PWN, Warsaw.

Czerwińska M., 2015, *The Specificity of E-Patients' Behaviour on the Internet*, "Roczniki" Kolegium Analiz Ekonomicznych, Oficyna Wydawnicza Szkoła Główna Handlowa, Warsaw, no. 38, 343–351.

Czyżewski L., Rybak K., 2018, *Assessment of the Exposure of Nurses from Emergency Departments to Aggressive Behaviour of Patients*, Disaster and Emergency Medicine Journal, vol. 3, no. 4: 111–118.

Davids J., Murphy M., Moore N., Wand T., Brown M., 2021. *Exploring Staff Experiences: A Case for Redesigning the Response to Aggression and Violence in the Emergency Department*. International Emergency Nursing, vol. 57.

Dawson D., Reid K., 1997, *Fatigue, Alcohol and Performance Impairment*, Nature, vol. 388, no. 6639: 235. doi: 10.1038/40775.

Deber R.B., Kraetschmer N., Urowitz S., Sharpe N., 2005, Dec, *Patient, Consumer, Client, or Customer: What Do People Want to Be Called?*, Health Expectations, vol. 8, no. 4: 345–351.

de Jong T., Bos E., Pawłowska-Cyprysiak K. et al., 2014, *Existing and Emerging Risks in the Healthcare Sector, Including Home and Non-Hospital Care, European Risk Observatory. Summary of the Report*, European Agency for Safety and Health at Work (EU-OSHA), Luxembourg.

de la Fuente M., Schoenfisch A., Wadsworth B. et al., 2019, *Impact of Behavior Management Training on Nurses' Confidence in Managing Patient Aggression*, Journal of Nursing Administration, vol. 49, no. 2: 73–78.

Dilts R., 1990, *Changing Belief Systems With NLP*, Dilts Strategy Group, Cupertino, Calif.

di Martino V., 2003, *Workplace Violence in the Health Sector. Relationship between Work Stress and Workplace Violence in the Health Sector*, Steering Committee of the ILO/ICN/WHO/PSI Joint Programme on Workplace Violence in the Health Sector, Geneva.

Directive 2000/54/EC of the European Parliament and of the Council on the protection of workers from risks related to exposure to biological agents at work. Official Journal of the European Communities, L 262, 17 October 2000.

Dodge R., Daly A.P., Huyton J., Sanders L.D., 2012, *The Challenge of Defining Wellbeing*, International Journal of Wellbeing, vol. 2, no. 3: 222–235. doi:10.5502/ijw.v2i3.4.

Dollard J., 1939, after: Kumięga P., 2014, *Theories of Aggression* [in:] A. Steciwko, J. Barański, A. Mastalerz-Migas (eds.), *Aggression in Healthcare. Introductory Issues*, Elsevier Urban &Partner, Wroclaw; Lickiewicz J., 2017, *The Aggressive Patient. Methods and Techniques of Management,* PZWL Medical Publishing House, Warsaw.

Dondajewska A., 2016, *Case Studies in Management Science Research in the Light of Methodological Rigor*, Scientific Journals of the Poznań University of Technology, No. 70, 39–50.

Dorosz G., Kowalczuk K., 2019, *The Impact of the Work Environment on the Occurrence of Stress in the Profession of Paramedic* [in:] K. Kowalczuk, E. Krajewska-Kułak (eds.), *Selected Health and Occupational Problems of Healthcare Workers*, vol. 1, Medical University of Białystok, Białystok.

Dudziak R., 2018, *Occupational Health and Safety*, 2nd edition, EDICON Education Development Center, Poznań.

Duhigg C., 2013, *The Power of Habit: Why We Do What We Do in Life and Business*, trans. M. Guzowska, PWN Publishing House, Warsaw.

Dyląg A., Rosiński J., 2008, Change Management in the Conditions of Post-merger Enterprise Integration, [in:] A. Herdan [ed.] Mergers, Acquisitions. Selected Aspects of Integration, UJ Publishing House, Kraków, pp. 105–126.

Dyne L.V., Ang S., Botero I.C., 2003, *Conceptualizing Employee Silence and Employee Voice as Multidimensional Constructs*, Journal of Management Studies, vol. 40, no. 6: 1359–1392.

Dzierżanowska-Fangrat K., Pawińska A., Semczuk K. (eds.), 2010, *Hand and Environmental Hygiene of Poles. Break the Chain of Infection*, Dettol and the Children's Health Center, Warsaw.

Ejdys J., Kobylińska U., Lulewicz-Sas A., 2012, *Integrated Quality, Environmental and Occupational Safety Management Systems*, Białystok University of Technology Publishing House, Białystok.

European Agency for Safety and Health at Work – EU-OSHA, 2014, Current and Emerging Issues in the Healthcare Sector, Including Home and Community Care, European Risk Observatory Report, Luxembourg.

European Commission (EC), 2011, *Health and Safety Risks in the Healthcare Sector. A Guide to Prevention and Good Practice*, Publications Office of the European Union, Luxembourg.

Fernandes C.M., Bouthillette F., Raboud J.M. et al. 1999, *Violence in the Emergency Department: A Survey of Health Care Workers*, Canadian Medical Association Journal, vol. 161, no. 10: 1245–1248.

Fiłończuk-Wieczorkowska H., Żukrowska M., 2014, *Examples of Aggressive Behavior and Ways of Dealing with Aggressive Patients* [in:] A. Steciwko, J. Barański, A. Mastalerz-Migas (eds.), *Aggression in Healthcare Introductory Issues*, Elsevier Urban & Partner, Wrocław.

Florek S., Siemiginowska P., 2020, *Aggression at Work*, Jagiellonian University, Jagiellonian Library, Kraków.

Foldspang L., Mark M., Rants L.L. et al., 2014, *Working Environment and Productivity. A Register-Based Analysis of Nordic Enterprises*, Nordic Council of Ministers, Denmark. doi: 10.6027/TN2014-546.

Freud S., 1927, after: Kowalczuk K., Jankowiak B., Krajewska-Kułak E. et al., 2011, Theories of Aggression-Truth or Myths? Nursing Problems, no. 19(3): 396–400.

Frydrysiak K., Grześkowiak M., Podlewski R., 2014, *Patient Aggression as a Threat in the Work of Emergency Medical Teams*, Anesthesiology and Emergency Medicine, vol. 8: 381–391.

Gacki-Smith J., Juarez A.M., Boyett L. et al., 2009, *Violence against Nurses Working in US Emergency Departments*, JONA, vol. 39, no. 7/8: 340–349.

Gałusza M., 2014, *Guide for Employers*, 3rd edition, Tarbonus Publishing House, Krakow-Tarnobrzeg.

Gawęda A., Śnieżek A., Serzysko B., 2018, *Job Satisfaction in the Opinion of Surveyed Nurses*, Nursing and Public Health, vol. 8, no. 4: 269–276.

Ghosh M., Twigg D., Kutzer Y., Towell-Barnard A., de Jong G., Dodds M. 2019. *The Validity and Utility of Violence Risk Assessment Tools to Predict Patient Violence in Acute Care Settings: An Integrative Literature Review*, International Journal of Mental Health Nursing, vol. 28, no. 6: 1248–1267. doi: 10.1111/inm.12645.

Gillespie G.L., Gates D.M., Miller M. et al., 2010, *Workplace Violence in Healthcare Settings: Risk Factors and Protective Strategies*, Rehabilitation Nursing Journal, vol. 35, no. 5: 177–184.

Grudzień D., Zurzycka P., Radzik T., 2015, Opinions of *M*edical *S*taff on *A*ggressive *B*ehavior of *P*atients, Polish Review of Health Sciences, vol. 4, no. 45.

Gulla B., 2020, *Anger, Aggression, and Violence in Selected Psychopathological Disorders*, Jagiellonian University, Jagiellonian Library, Krakow.

Harms-Ringdahl L., 2004, *Relationships between Accident Investigations, Risk Analysis and Safety Management*, Journal of Hazardous Materials, vol. 111: 13–19.

Harvey J., Bolam H., Erdos G. et al., 2001, *The Effectiveness of Training to Change Safety Culture and Attitudes within a Highly Regulated Environment*, Personnel Review, vol. 30, no. 6: 615–636.

Health and Safety Executive (HSE), 2004, *Investigating Accidents and Incidents, A Workbook for Employers, Unions, Safety Representatives and Safety Professionals*, HSG245 02/14.

Heckemann B., Zeller A., Hahn S. et al., 2015, *The Effect of Aggression Management Training Programs for Nursing Staff and Students Working in an Acute Hospital Setting. A Narrative Review of Current Literature*, Nurse Education Today, vol. 35, no. 1: 212–219.

Helios J., Jedlecka W., 2017, *Contemporary Faces of Violence. Selected Issues*, E-Publishing House. Legal and Economic Digital Library. Faculty of Law, Administration and Economics, University of Wrocław, Wrocław.

Henriksen K., Dayton E., 2006, *Organizational Silence and Hidden Threats to Patient Safety*, Health Services Reasearch, vol. 41, no. 4, Pt 2: 1539–1554.

Hersey P., Blanchard K.H., Johnson D.E., 2000, *Management of Organizational Behavior: Leading Human Resources*, 8th edition, Prentice Hall, United States of America.

Ilczak T., Rak M., Ćwiertnia M. et al., 2021, *Predictors of Stress among Emergency Medical Personnel during the COVID-19 Pandemic*, International Journal of Occupational Medicine and Environmental Health, vol. 34, no. 2: 139–149.

ILO (International Labour Office), 2017, *Psychosocial Risks, Stress and Violence in the World of Work*, International Journal of Labour Research, Geneva, vol. 8, no. 1–2: 35–65.

ILO (International Labour Office), 2019, The Future of Work in the Health Sector, Geneva, Working Paper no. 325, 1–25.

Ipek M., Özlük B., 2023, *Patients' Opinion on Violence against Healthcare Workers and Their Level of Satisfaction in Emergency Departments in Turkey: A Cross-Sectional Study*, International Emergency Nursing, vol. 71: 101350. doi: 10.1016/j.ienj.2023.101350

Jasińska J., 2010, *Working Relations and Conditions* [in:] Listwan T. (ed.), *Human Resource Management*, C.H.Beck Publishing House, Warsaw.

Jaworek M., 2021, *The Affective Dimension of Work-Related Well-Being and Its Specificity in the Group of Managers*, Jagiellonian University Press, Krakow.

Juchnowicz M., Kinowska H., 2022, *Employee Components in Hybrid Working Conditions* [in:] J. Tabor-Błażewicz, H. Rachoń (eds.), *Challenges of Managing People in a Hybrid Work Organization System*, SGH Publishing House, Warsaw.

Kaeser D., Guerra R., Keidar O. et al., 2018, *Verbal and Non-Verbal Aggression in a Swiss University Emergency Room. A Descriptive Study*, International Journal of Environmental Research and Public Health, vol. 15: 1423.

Kenrick D.T., Neuberg S.L., Cialdini R.B., 2002, *Social Psychology*, Gdańsk Psychological Publishing House, Gdańsk.

Kim Y., Park J., Park M., 2016, *Creating a Culture of Prevention in Occupational Safety and Health Practice*, Safety and Health at Work, vol. 7, no. 2: 89–96.

Kirwan B., Reader T., Parand A., 2019, *The Safety Culture Stack – the Next Evolution of Safety Culture?*, Safety and Reliability, vol. 38, no. 3: 200–217.

Kmiecik B., 2016, *The Problem of Patient Aggression – an Attempt at Diagnosis, Search for Solutions*, Neurological News, vol. 16, no. 3: 158–162.

Kmiecik-Baran K., 1998, *Violence against Children – Diagnosis and Intervention* [in:] J. Papież, A. Płukis (eds.), *Violence against Children and Young People in the Context of Poland's Political Transformation*, Wydawnictwo Adam Marszałek, Toruń.

Kowalczuk K., Krajewska-Kułak E., 2017, *Patient Aggression towards Different Professional Groups of Healthcare Workers*, Annals of Agricultural and Environmental Medicine, vol. 24, no. 1: 113–116.

Kowalski J., 1997, *Legal Basis for Occupational Safety in Poland* [in:] D. Koradecka (ed.), *Occupational Safety and Ergonomics*, vol. 2, Central Institute for Labor Protection, Warsaw.

Krahé B., 2005, *Aggression*, trans. J. Suchecki, Gdańsk Psychological Publishing House, Sopot.

Krot K., 2008, *Quality and Marketing of Medical Services*, ABC and Wolters Kluwer Business, Warsaw.

Krug E.G., Dahlberg L.L., Mercy J.A. et al., 2002, *World Report on Violence and Health*, World Health Organization, Geneva.

Kudryńska B., Krzywdziński K., 2021, *The Impact of Belonging to the Professional Group of Nurses and Doctors on Organizational Silence*, Public Health and Management, vol. 19, no. 3–4: 123–129.

Kulig-Moskwa K., 2018, *The Essence of Wellbeing in an Organization* [in:] K. Kulig-Moskwa (ed.), *Wellbeing in an Organization. What? How? Why?*, Infor Publishing House, Warsaw.

Kumięga P., 2014, *Theories of Aggression* [in:] A. Steciwko, J. Barański, A. Mastalerz-Migas (eds.), *Aggression in Healthcare. Introductory Issues*, Elsevier Urban & Partner, Wrocław.

Kuriata-Kościelniak E., 2020, *The Importance of Interpersonal Communication for the Safety of Patients and Medical Staff* [in:] I. Witczak, Ł. Rypicz (eds.), *Safety of Patients and Medical Staff. Ergonomic Conditions*, Medical University of Silesia Press, Wrocław.

Labor Market Analysis Institute, 2020, Analysis of the Employment and Wage Gap between Women and Men, Report, Warsaw.

Lee H., Yun H., Choi M., Kim H., 2023, *Predicting Workplace Violence in the Emergency Department Based on Electronic Health Record Data*, Journal of Emergency Nursing, vol. 49, no. 3: 415–424.

Leka S., Jain A., 2013, *Psychosocial Hazards in the Work Environment and their Impact on Health*, translated by Katarzyna Orlak, Healthy Work Association, Warsaw.

Lewin K., 1947, Group Decisions and Social Change, [in:] T.M. Newcomb E.L. Hartley (eds.) 1959. *Reading in Social Psychology*, Henry Holt, New York.

Lickiewicz J., 2017, *Aggressive Patients. Methods and Techniques for Dealing with Them*, PZWL Medical Publishing House, Warsaw.

Lickiewicz J., Marzęcka A., Makra-Studzińska M., 2018, *Comparative Analysis of Aggressive Behavior in the Emergency Room and Surgery Ward – Preliminary Report*, Anesthesiology and Emergency Medicine, vol. 12: 266–273.

Lickiewicz J., Piątek J., 2014, *Experience of Aggression in Nursing Work*, Sztuka Leczenia, no. 3–4: 11–12.

Liwo M.A., 2013, Occupational Safety (Work Environment) and First Aid and Emergency Medical Services, Emergency Room Occupational Safety (Work Environment) and First Aid and Emergency Medical Services, Emergency Room, v6, no 1: 17–24.

Lorenz K., 1963, after: Kowalczuk K., Jankowiak B., Krajewska-Kułak E. et al., 2011, *Theories of Aggression-Truth or Myths?* Nursing Problems, no. 19(3):396–400; Kumięga P., 2014, *Theories of Aggression* [in:] A. Steciwko, J. Barański, A. Mastalerz-Migas (eds.), *Aggression in Healthcare. Introductory Issues,* Elsevier Urban &Partner, Wrocław.

Markiewicz R., 2012, *Aggressive Behavior towards Nursing Staff Employed in Psychiatric Wards*, Current Problems of Psychiatry, vol. 13, no. 2: 93–97.

Martinez A., 2016, *Managing Workplace Violence with Evidence-Based Interventions: A Literature Review*, Journal of Psychosocial Nursing and Mental Health Services, vol. 54, 31–36.

Mazur J., 2002, *Domestic Violence. Theory and Reality*, Wydawnictwo Akademickie Żak, Warsaw.

McNamara R., Yu D.K., Kelly J.J., 1997, *Public Perception of Safety and Metal Detectors in an Urban Emergency Department*, The American Journal of Emergency Medicine, vol. 15, no. 1: 40–42.

Merecz D., Mościcka A., 2003, Violence in Healthcare Facilities and Ways to Counteract It, Prof. J. Nofer Institute of Occupational Medicine in Łódź, Łódź.

Meyer T., Wrenn K., Wright S.W., Glaser J., Slovis C.M., 1997, *Attitudes toward the Use of a Metal Detector in an Urban Emergency Department*, Annals of Emergency Medicine, vol. 29, no. 5: 621–624.

Mikos M. (ed.), 2017, *Patient Safety*, Ziz Education Center, Krakow.

Mikos M., Kopacz P., Żukrowska M., 2018, *The Issue of Aggression towards Paramedics in Emergency Medical Teams*, Safety Theory and Practice, no. 2, 57–65.

Miller N.E., 1950, after: Kumięga P., 2014, *Theories of Aggression* [in:] A. Steciwko, J. Barański, A. Mastalerz-Migas (eds.), *Aggression in Healthcare. Introductory Issues*, Elsevier Urban &Partner, Wrocław.

Morrison E.W., Milliken F.J., 2000, *Organizational Silence: A Barrier to Change and Development in a Pluralistic World*, Academy of Management Review, vol. 25, no. 4: 706–725.

Mościcka A., Drabek M., 2010, *Individual and Environmental Factors Contributing to Exposure to Mobbing*, Occupational Medicine, vol. 61, no. 4.

Muraven M., Gagné M., Rosman H., 2008, *Helpful Self-Control: Autonomy Support, Vitality, and Depletion*. Journal of Experimental Social Psychology, vol. 44, no. 3: 573–585. https://doi.org/10.1016/j.jesp.2007.10.008.

NHS, 2020, *Violence Prevention and Reduction Standard*, NHS England and NHS Improvement, London.

Nielsen K.J., 2014, *Improving Safety Culture through the Health and Safety Organization: A Case Study*, Journal of Safety Research, vol. 48: 7–17.

Nieva V.F., Sorra J., 2003, *Safety Culture Assessment: A Tool for Improving Patient Safety in Healthcare Organizations*, Quality & Safety in Health Care, Suppl. 2, Suppl. II: ii17–ii23.

O'dea A., Flin R., 2001, *Site Managers and Safety Leadership in the Offshore Oil and Gas Industry*, Safety Science, vol. 37, 39–57.

Okundolor S., Ahenkorah F., Sarff L. et.al., 2021, *Zero Staff Assaults in the Psychiatric Emergency Room: Impact of a Multifaceted Performance Improvement Project*, Journal of American Psychiatric Nurses Association, vol. 27, 64–71.

Pacana A., Ostasz G., Olejarz T., 2017, *Updated Approach to the Methodology of Implementing the Occupational Health and Safety Management System* [in:] A. Pacana, G. Ostasz (eds.), *Hazards in the Workplace. Selected Issues*, Rzeszów University of Technology Publishing House, Rzeszów.

Pallarés J.M.C., Ruiz-Hernández J.A., Galián-Muñoz I. et al., 2023, *Workplace Violence in Hospital Emergency Departments and Consequences for Health Care Professionals and Support Staff: A Cluster – Al Analysis*, Emergencias, vol. 35, no. 5: 353–358. doi: 10.55633/s3me/E026.2023.

Parker D., Lawrie M., Hudson P., 2006, *A Framework for Understanding the Development of Organisational Safety Culture*, Safety Science, vol. 44, no. 6: 551–562.

Perrone S., 1999, *Violence in the Workplace*, Australian Institute of Criminology, Canberra, Australia.

Pilszyk A., 2007, *Psychopathological Picture of the Perpetrator of Domestic Violence*. Polish Psychiatry, vol. 51, no. 6: 827836.

Piotrowski A., Nikkhhah-Farkhani Z., Makarowski R., 2021, *Stress Among Nurses from Countries Variously Affected by the COVID-19 Pandemic – Results Collected in Poland and Iran*, Occupational Medicine, vol. 72, no. 5: 457–465.

Płaszewska-Żywko L., Podstawa M., Milan H. et al., 2019, *Occurrence of Patient's Aggressive Behaviours towards Nurses and Paramedics*, Nursing Problems, vol. 27, no. 3–4: 144–148.

Rabin S., Akinfemiwa O., Bradley M. et al., 2025, *Protecting Frontline Workers: Strategies for Preventing and Mitigating Violence in the Emergency Department*, Annals of Emergency Medicine, vol. 85, no. 3: 263–269. doi: 10.1016/j.annemergmed.2024.06.022.

Rączkowski B., 2018, *Initial Occupational Health and Safety Training. General Instruction*, 5th edition, Center for Consulting and Staff Development, Gdańsk.

Rączkowski B., 2020, *Health and Safety in Practice. A Guide for Health and Safety Service Employees, Employers, Labor Inspectors, Social Labor Inspectors, Designers, Lecturers*, 18th edition, Center for Consulting and Staff Development, Gdańsk.

Ramacciati N., Gili A., Mezzetti A. et al., 2019, *Violence towards Emergency Nurses: The 2016 Italian National Survey – A Cross-Sectional Study*, Journal of Nursing Management vol. 27, no. 4: 792–80. doi: 10.1111/jonm.12733.

Ratnapalan S., 2009, *Shades of Grey: Patient Versus Client*, Canadian Medical Association Journal, vol. 180, no. 4: 472.

Richardson S.K., Ardagh M.W., Morrison R. et al., 2019, *Management of the Aggressive Emergency Department Patient: Non-Pharmacological Perspectives and Evidence Base*, Open Access Emergency Medicine, vol. 11, 271–290.

Roche M.A., Diers D., Duffield C. et al., 2010, *Violence toward Nurses, the Work Environment and Patient Outcomes*, Journal of Nursing Scholarship, vol. 42, no. 1: 13–22.

Rode D., 2010, *Psychological Determinants of Domestic Violence. Characteristics of Perpetrators*, University of Silesia Press, Katowice.

Rosiński J., Filipkowska A., 2009, *Knowledge Workers – How to Develop Specialist Competences for the Benefit of the Employee and the Organization* [in:] E. Skrzypek, A. Sokół (eds.), *Human Capital Management in a Knowledge-Based Economy*, Institute of Knowledge and Innovation, Warsaw, pp. 213–228.

Rosiński J., Różańska A., Jarynowski A. et al., 2019, *Factors Shaping Attitudes of Medical Staff towards Acceptance of the Standard Precautions*, International Journal of Environmental Research and Public Health, vol. 16, no. 6: 1050.

Rundmo T., Hale A.R., 2003, *Managers' Attitudes towards Safety and Accident Prevention*, Safety Science, vol. 41, 557–574.

Rypicz Ł., 2020, *Psychosocial Risk Factors in the Work Environment of Medical Staff and Patient Safety* [in:] I. Witczak, Ł. Rypicz (eds.), *Safety of Patients and Medical Staff. Ergonomic Conditions*, Piastów Śląskich Medical University in Wrocław, Wrocław.

Sachdeva S., Jamshed N., Aggarwal P., Kashyap S.R., 2019, *Perception of Workplace Violence in the Emergency Department*, Journal of Emergencies, Trauma, and Shock, vol. 12, no. 3: 179–184.

Sidor-Rządkowska M., 2018, *Human Resource Management in Public Hospitals – Problems and Challenges*, Studies and Papers. College of Management and Finance, vol. 167: 127–141.

Sinclair, R. R., Sears, L. E., Probst, T., & Zajack, M. (2010). A multilevel model of economic stress and employee well-being. In J. Houdmont & S. Leka (Eds.), *Contemporary occupational health psychology: Global perspectives on research and practice,* Vol. 1, pp. 1–20. Wiley Blackwell. https://doi.org/10.1002/9780470661550.ch1

Sharifi S., Shahoei R., Nouri B., Almvik R., Valiee S., 2020, *Effect of an Education Program, Risk Assessment Checklist and Prevention Protocol on Violence against Emergency Department Nurses: A Single Center Before and After Study*, International Emergency Nursing, vol. 50, no. 100813. doi: 10.1016/j.ienj.2019.100813.

Sheldon W.H., 1940, after: Kumięga P., 2014, *Theories of Aggression* [in:] A. Steciwko, J. Barański, A. Mastalerz-Migas (eds.), *Aggression in Healthcare. Introductory Issues*, Elsevier Urban &Partner, Wrocław.

Sommer H., 2013, *Electronic Aggression – a Few Remarks on the Phenomenon* [in:] K. Barłóg, E. Tłuczek-Tadla (eds.), *Aggression and Violence, Selected Contexts and Challenges*. Pedagogical Discourse, Publishing House of the Bronisław Markiewicz State Higher School of Technology and Economics in Jarosław, Jarosław.

Somville F., Vanspringel E., De Cauwer H. et al., 2021, *Work Stress-Related Problems in Physicians in the Time of COVID-19*, International Journal of Occupational Medicine and Environmental Health, vol. 34, no. 3: 373–383.

Sorensen G., McLellan D.L., Sabbath E.L., Dennerlein J.T., Nagler E.M., Hurtado D.A., Pronk N.P., Wagner G.R., 2016, *Integrating Worksite Health Protection and Health Promotion: A Conceptual Model for Intervention and Research*, Preventive Medicine, 91, 188–196. https://doi.org/10.1016/j.ypmed.2016.08.005.

Stańczyk I., 2012, *Rationalization and Restructuring of Employment in Healthcare* [in:] M. Bugdol, J. Bugaj, I. Stańczyk (eds.), *Human Resource Management Processes in Healthcare*, State Higher Vocational School in Opole, Continuo Publishing House, Wrocław.

Stankiewicz M., Sznajder M., 2010, *Occupational Health and Safety Culture in Organizations* [in:] J. Ejdys (ed.), *Shaping Occupational Health and Safety Culture in Organizations*, Białystok University of Technology Publishing House, Białystok.

Stansfield S.A., Matheson M.P., 2003, *Noise Pollution: Non-Auditory Effects on Health*, British Medical Bulletin, vol. 68: 243–257.

Staszkiewicz M., Klimkiewicz K., Beck-Krala E., 2023, *Designing and Implementing Programs Focused on Employee Well-Being in Poland* [in:] M. Juchnowicz, H. Kinowska (eds.) *Human Capital Management in Conditions of Uncertainty. Challenges and Implications*, Polish Economic Publishing House, Warsaw.

Stępień A., 2015, The Phenomenon of Aggression in the youth Environment. The Concept of Aggression, Violence and their Symptoms, Family Pedagogy. Pedagogy of the Family, vol. 5, no. 3.

Suwała A., 2008, *The Phenomenon of Violence in Healthcare Facilities*, Nursing Problems, vol. 16, no. 1–2: 157–162.

Szkup M., Rączka M., Schneider-Matyka D. et al., 2018, *Assessment of the Risk of Aggression among Patients Visiting Medical Facilities*, Public Health and Management, No, vol. 16, no. 2: 119–124.

Szłązak J., Szłązak N., 2012, *Occupational Health and Safety*, 3rd edition, AGH Publishing House, Krakow.

Szwamel K., Sochocka L., 2014, *Opinion of Mid-Level Medical Staff in Hospital Emergency Departments on Aggressive Patient Behavior*, Nursing and Public Health, vol. 4, no. 2: 149–154.

Szydłowska D., Jankowiak B., 2019, *The Degree of Nurses' Exposure to Aggression in the Workplace* [in:] Kowalczuk K., Krajewska-Kułak E. (eds.), *Selected Health and Occupational Problems of Healthcare Workers*, vol. 1, Medical University of Białystok, Białystok.

Tabor-Błażewicz J., 2022, *The Impact of Remote and Hybrid Work on Employee Well-Being* [in:] J. Tabor-Błażewicz, H. Rachoń (eds.), *Challenges of Managing People in a Hybrid Work Organization System*, SGH Publishing House, Warsaw.

Tarczoń E., Beck-Krala E., Kowalik W., 2025, *Medical Staff Facing Aggressive Behavior from Patients. Challenges for Management*, AGH Publishing House, Krakow.

Tobiasz-Adamczyk B., 2014, *Selected Aspects of Aggression and Violence from a Sociological Perspective* [in:] A. Steciwko, J. Barański, A. Mastalerz-Migas (eds.), *Aggression in Healthcare Introductory Issues*, Elsevier Urban & Partner, Wrocław.

Tung L.T, Anh T.T, Chinh N.T.M, Long N.H., 2021, *Public Reactions in Online Newspapers to Workplace Violence against Nurses*. Sage Open Nursing, vol. 23, no. 7. doi: 10.1177/23779608211003819.

Uzarczyk A., 2009, *Harmful and Burdensome Factors in the Work Environment*, 2nd edition, Center for Consulting and Staff Development, Gdańsk.

Van den Tooren M., De Jonge J., 2010, *The Role of Matching Job Resources in Different Demanding Situations at Work: A Vignette Study*, Journal of Occupational and Organizational Psychology, vol. 83, no. 1: 39–54.

WHO & ILO (World Health Organization and International Labour Organization), 2018, *Occupational Safety and Health in Public Health Emergencies*, Geneva.

Wieczorek S., 2014, *Ergonomics*, 3rd edition, Tarbonus Publishing House, Kraków-Tarnobrzeg.

Wieczorek S., Żukowski P., 2014, *Organization of Safe Work*, Tarbonus Publishing House, Kraków-Tarnobrzeg.

Wiskow C., Albreht T., de Pietro C., 2010, *How to Create an Attractive and Supportive Working Environment for Health Professionals*, World Health Organization on Behalf of the European Observatory on Health Systems and Policies, Publications WHO Regional Office for Europe, Denmark.

Wiśniewska M.Z. (ed.), 2016, *Quality of Medical Services. Instruments and Models*, CeDeWu.pl Publishing House, Warsaw.

Wiśniewska M.Z., 2021, *Organizational Culture and Cultures That Strengthen the Improvement of Healthcare Entities*, Spatium Scientific and Publishing Institute, Radom.

Witczak I., Rypicz Ł. (eds.), 2020, *Safety of Patients and Medical Personnel. Ergonomic Conditions*, Publishing House of the Piastów Śląskich Medical University in Wrocław, Wrocław.

Wróbel A., 2014, *Occupational Health and Safety – Concept and Responsibility*, Review of Law and Administration, Wrocław, no. 96: 93–100.

Yang B.X., Stone T.E., Petrini M.A. et al., 2018, *Incidence, Type, Related Factors, and Effect of Workplace Violence on Mental Health Nurses: A Cross-Sectional Survey*, Archives of Psychiatric Nursing, vol. 32, no. 1: 31–38. doi: 10.1016/j.apnu.2017.09.013.

Ziaei M., Massoudifar A., Rajabpour-Sanati A. et al., 2019, *Management of Violence and Aggression in Emergency Environment; a Narrative Review of 200 Related Articles*, Journal of Emergency Medicine, vol. 3, no. 1: e7.

Żurowska-Wolak M., Wolak B., Mikos M. et al., 2015, *Stress and Burnout in the Work of Paramedics*, Journal of Education, Health and Sport, vol. 5, no. 7: 43–50.

Zużewicz K., 2017, *Health Effects of Working in a Non-Physiological Rhythm*, Scientific Journals of the Main School of Fire Service, Warsaw, no. 62: 127–139.

Netography

1. *A Declaration on the Promotion of Patients' Rights in Europe,* Europe Consultation on the Rights of Patients, Amsterdam 28–30 March 1994 (https://www.activecitizenship.net/multimedia/files/charter-of-rights/the-forerunners-of-the-charter/The-Declaration-on-The-Promotion-of-Patients-Rights-in-Europe.pdf accessed on: 17.04.2024).
2. *Reducing Occupational Violence and Aggression in Victorian Health Services,* (https://www.health.vic.gov.au/worker-health-wellbeing/reducing-occupational-violence-and-aggression-in-victorian-health-services accessed on: 17.11.2023).
3. *A European Health Union. A Blueprint for Generations* 2023, Edited by: Vytenis Povilas Andriukaitis and Gediminas Cerniauskas, (https://feps-europe.eu/wp-content/uploads/2023/09/Book_A-European-Health-Union-2.pdf accessed on: 24.11.2024).
4. *APA Resolution for the Use of the Term Patient in American Psychological Association Policies, Rules, and Public Relations Activities when Referring to the Health-Related and Scientific Activities of Health Service Psychologists and Scientists in Health Care Services and Settings,* August 2018 (https://www.apa.org/about/policy/resolution-term-patient.pdf accessed on: 26.03.2024).
5. CIOP-BIP, *Dangerous, Harmful, and Burdensome Factors and Their Consequences in Micro and Small Retail Businesses* (https://m.ciop.pl/CIOPPortalWAR/appmanager/ciop/mobi?_nfpb=true&_pageLabel=P4180015671496742791029&html_tresc_root_id=300007035&html_tresc_id=300007030&html_klucz=300006222&html_klucz_spis accessed on: 27.04.2021).
6. Cewińska 2016, *Fear in the Workplace and Its Negative Consequences.* (https://www.researchgate.net/publication/308702404_Fear_in_the_workplace_and_its_negative_consequences_Fear_in_the_workplace_and_its_negative_consequences#fullTextFileContent accessed on: 07.07.2024).
7. Costa D., Mercieca-Bebber R., Tesson S. et al. (2019), *Patient, Client, Consumer, Survivor or Other Alternatives? A Scoping Review of Preferred Terms for Labelling Individuals Who Access Healthcare across Settings,* BMJ Open (https://bmjopen.bmj.com/content/bmjopen/9/3/e025166.full.pdf accessed on: 23.03.2024).
8. Department of Health Victoria in Australia. https://www.health.vic.gov.au/worker-health-wellbeing/occupational-violence-and-aggression-health-facility-design accessed on: 02.01.2026.
9. Health Policy.com., *During the Pandemic there has been an Increase in Aggression in Hospitals and Clinics,* (https://politykazdrowotna.com/artykul/w-czasie-pandemii-przybylo-agresji-w-szpitalach-i-przychodniach-n824959 accessed on: 13.03.2022).

10. *Emergency Care in Victoria* (https://www.health.vic.gov.au/patient-care/addressing-violence-in-victorian-healthcare-settings accessed on: 27.11.2023).
11. EU-OSHA 2021, *High Levels of Occupational Safety and Health Pay Off for Businesses* (https://osha.europa.eu/pl/themes/good-osh-is-good-for-business accessed on: 17.03.2022).
12. *European Emergency Medicine in Numbers. Emergency Medicine. Epidemiology Series* (https://eusem.org/images/European_EM_in_numbers.pdf accessed on: 24.11.2024).
13. EWCS – *European Working Conditions Survey* (https://www.eurofound.europa.eu/en/topic/living-and-working-europe accessed on: 27.11.2023).
14. Grudzień, Zurzycka, Radzik, 2015, *Opinions of Medical Staff on Aggressive Patient Behavior,* Polish Review of Health Sciences, 4 (45) : 242-247 https://przeglad.ump.edu.pl/uploads/2015/4/242_4_45_2015.pdf accesssed on: 21.11.2022.
15. *Health at a Glance: Europe 2024, State of Health in The EU Cycle* (https://www.oecd.org/en/publications/health-at-a-glance-europe-2024_b3704e14-en/full-report.html accessed on: 24.11.2024).
16. Le T., Himmelstein D., Greene C., 2021, *greenelab/iscb-diversity: ISCB analysis v4.0 release (Zenodo)* (https://doi.org/10.5281/zenodo.4672417 accessed on: 27.10.2024
17. Lee H., Yun H., Choi M., Kim H. 2023, *Predicting Workplace Violence in the Emergency Department Based on Electronic Health Record Data,* Journal of Emergency Nursing, Volume 49, Issue 3, Pages 415-424, https://doi.org/10.1016/j.jen.2023.01.010 accessed on: 21.07.2024
18. Najmeh Sharifi, Mohsen Adib-Hajbaghery, Maryam Najafi, 2019, *Cultural Competence in Nursing: A Concept Analysis,* International Journal of Nursing Studies, Volume 99,103386, ISSN 0020-7489, https://doi.org/10.1016/j.ijnurstu.2019.103386 accessed on: 27.07.2024
19. NIOSH (National Institute for Occupational Safety and Health), Department of Health and Human Services Centers for Disease Control and Prevention, 2008, *Exposure to Stress,* Occupational Hazards in Hospitals (https://www.cdc.gov/niosh/docs/2008-136/pdfs/2008-136.pdf accessed on: 07.05.2021).
20. NIOSH 2023, *Total Worker Health Program,* Centers for Disease Control and Prevention (https://www.cdc.gov/niosh/twh/default.html accessed on: 23.03.2024).
21. "*Occupational Violence Prevention in Queensland Health's Hospital and Health Services, Taskforce Report,* State of Queensland (Queensland Health), May 2016". https://www.health.qld.gov.au/__data/assets/pdf_file/0024/443265/occupational-violence-may2016.pdf accessed on: 2.01.2026.
22. PTLECHZ – Polish Society of Epidemiologists and Infectious Disease Physicians (https://www.pfed.org.pl/uploads/1/9/9/8/19983953/raport_zon_final.pdf accessed on: 17.07.2022).
23. EU-OSHA Report 2023, *Occupational Safety and Health in Europe: Status and Trends 2023* (https://osha.europa.eu/sites/default/files/OSH_in_Europe_state_trends_report_2023_en.pdf accessed on: 13.12.2023).

24. Simińska A., 2016, *Aggressive and Vulgar Patient? Write It Down in the Documentation* (https://orbidenti.com/a/124/agresywny-i-wulgarny-pacjent-napisz-to-w-dokumentacji accessed on: 07.04.2022).
25. *10 Differences between a Client and a Patient* (2024) (https://www.ambula.io/10-differences-between-a-client-and-a-patient/ accessed on: 22.03.2024).
26. *The Royal Melbourne Hospital* (https://www.thermh.org.au/your-care/hospital-care/what-you-need-to-know-up-front/aggressive-violent-behaviour accessed on: 17.11.2023).
27. Trossman S., 2007, *Too Tired to Be Safe? ANA Position Statements Call for RNs and Employers to Prevent Working Fatigued* (https://www.americannursetoday.com/tootiredtobesafeanapositionstatements-call-for-rns-and-employers-to-preventworking-fatigued/ accessed on: 13.10.2019).
28. "*Well-Being at Work: Innovation and Good* Practice", European Working Conditions Observatory, 2011 (https://www.eurofound.europa.eu/ewco/2011/06/FI1106011I.htm, accessed on: accessed on: 18.03.2023).
29. *NHS Violence Prevention and Reduction Standard.* Guidance Notes. https://www.england.nhs.uk/wp-content/uploads/2022/06/B0989-NHS-violence-prevention-and-reduction-standard-guidance-notes.pdf accessed on: 13.10.2024
30. *About Workplace Violence* (https://www.cdc.gov/niosh/violence/about/ accessed accessed on:13.04.2024).
31. *Developing Healthcare Safety & Violence Prevention Programs within Hospitals*, https://www.aha.org/system/files/media/file/2019/08/mha-developing-healthcare-safety-violence-prevention-programs-in-hospitals.pdf accessed on: 12.04.2025
32. World Health Organization, 2002, (https://iris.who.int/server/api/core/bitstreams/a25476ed-8585-47f3-986e-9d0e7f5e9f1b/content accessed on: 13.07.2021).

GLOSSARY

ABSENCE a general term for not being where one should be, most often at work.

AGGRESSION behavior intended to cause someone harm, pain, damage, or suffering, whether physical, psychological, or symbolic.

ANESTHETIC GAS an inhaled anesthetic administered as primary treatment for preoperative sedation and as an adjunct to maintain intravenous anesthesia.

CONCEPTUAL ERGONOMICS deals with the design of workstations so that they ensure maximum safety at work with the least possible physical and mental strain on the employee.

CORRECTIVE ERGONOMICS deals with the analysis of existing working conditions in terms of their adaptation to the psychophysical capabilities of employees and the formulation of recommendations for improving the material factors of the working environment.

CUSTOMER a natural person or company that purchases goods or services to meet their needs in exchange for a fee.

CYTOSTATIC a drug used in the treatment of cancer, which inhibits the proliferation of cancer cells by damaging their DNA or disrupting the cell division process.

DANGEROUS FACTOR a factor whose effect may lead to immediate deterioration of health, accidents at work, or injuries.

MEDICAL RESCUE TEAM paramedics, nursing staff, and doctors who provide medical assistance to people in a state of sudden health emergency outside of a hospital.

ERGONOMICS the science of finding the best ways and means to ensure optimal conditions in the workplace.

HARMFUL FACTOR a factor whose effect may lead to a gradual deterioration in health and then to illness.

HEALTH SERVICES activities aimed at preserving, saving, restoring, and improving health.

HOSPITAL EMERGENCY DEPARTMENT (ED) a separate part of the hospital, operating 24 hours a day, equipped to diagnose and treat patients in emergency situations.

IMMATERIAL FACTOR a social, organizational, and psychological element that affects the employee's experience and is not related to the physical work space.

ISO 45001:2024 an international standard specifying requirements for an occupational health and safety management system aimed at preventing accidents and occupational diseases and continuously improving working conditions.

LONG-TERM FATIGUE SYNDROME a disorder caused by disruption of the circadian rhythm, manifested by sleep problems, fatigue, irritability, headaches, appetite, and concentration disorders.

MATERIAL FACTOR an element of the environment directly related to an accident or deviation from the normal state, which may cause injury or contribute to an incident.

MEDICAL ERROR an action or omission by a medical professional that is inconsistent with current medical knowledge and leads to unintentional injury or death of a patient.

MEDICAL SERVICES elements of health services extended to include activities aimed at maintaining the good mental, physical, and social development of people.

MENTAL STRAIN excessive mental and emotional exertion caused by the accumulation of tasks, stress, and responsibilities.

MICROORGANISM a microscopic organism that includes bacteria, fungi, protozoa, and algae.

NUISANCE FACTOR a factor whose effect may reduce the ability to perform work and decrease productivity, but does not cause permanent deterioration in health.

OCCUPATIONAL HEALTH AND SAFETY a science concerned with studying the impact of working conditions on employee health and preventing the negative effects of work.

OCCUPATIONAL HEALTH AND SAFETY MANAGEMENT a method of occupational risk management aimed at reducing the risk of loss of life and health at work to an acceptable level.

OCCUPATIONAL RISK the likelihood of undesirable work-related events occurring that may lead to a deterioration in the employee's health or to material losses.

OCCUPATIONAL RISK ASSESSMENT a method leading to improved occupational health and safety, which provides feedback to the employer on compliance with legal requirements in the field of occupational safety.

OCCUPATIONAL STRESS harmful physical and emotional reactions that occur when job demands exceed an employee's capabilities.

ORGANIZATIONAL SILENCE giving up the possibility of influencing a situation that is considered undesirable, manifested by a lack of faith in the possibility of change and corrective action.

PATIENT a person using healthcare services, regardless of whether they are healthy or sick.

PERSONAL PROTECTIVE EQUIPMENT any device, equipment, or accessory designed to be worn or held by an employee to protect them from hazards in the work environment.

PERSONNEL FUNCTION a set of activities of an organization related to people, aimed at achieving the company's goals and satisfying the needs of employees.

PHYSICAL STRAIN a set of external forces acting on the body that require muscular effort.

PHYSICAL STRESS the system of external forces acting on the body that require muscular effort.

PHYSICAL VIOLENCE a violation of physical integrity that causes pain or bodily harm.

PROFESSIONAL PRIDE a sense of satisfaction and value from one's work, which results from recognizing the importance of one's actions and a sense of excellence in their implementation.

PSYCHOLOGICAL VIOLENCE situations in which the perpetrator exercises psychological control over the victim.

PSYCHOPHYSICAL FACTOR a factor that causes physical strain, e.g., work overload.

PSYCHOSOCIAL FACTOR a factor that causes mental strain.

PUBLIC OFFICIAL a person who performs a public function under the law.

QUALITY OF MEDICAL SERVICES a comprehensive activity in which the extent to which a medical service meets the needs and expectations of the patient is assessed, covering factors such as safety, effectiveness, efficiency, timeliness, and patient-centeredness.

SAFETY CULTURE the result of individual and group attitudes, values, perceptions, competencies, behavior patterns, and the style and quality of safety management in an organization.

SAFETY MANAGEMENT SYSTEM a set of organizational measures aimed at ensuring conditions that allow work to be performed safely, without harm to health.

SHIFT WORK a work organization system in which employees in the same position change at fixed times, allowing the organization to operate continuously around the clock.

SOCIALLY UNACCEPTABLE BEHAVIOR any actions that violate the norms, values, and principles applicable in a given group, leading to conflicts with others and hindering the normal functioning of the individual.

STERILIZATION the process of destroying all microorganisms, including their spore forms.

SUBSTANTIVE COMPETENCE specialist knowledge and skills relating to a given field or matter.

SUBSTITUTE a replacement, product, or service that can satisfy the same needs as another, but differs in composition or method of manufacture.

TIME PRESSURE a feeling of compulsion caused by an approaching deadline for completing tasks, a subjective impression that can lead to stress, a feeling of helplessness, and a decline in the quality of work.

TRIAGE a medical procedure involving the rapid assessment of patients' health and their segregation in order to determine priorities in providing assistance.

TOURETTE'S SYNDROME a neurological disorder characterized by tics, i.e., sudden, involuntary movements and sounds.

VIOLENCE an intentional act that violates the rights and personal interests of another person, threatens their life, health, dignity, physical integrity, freedom, or causes suffering.

WELL-BEING a relative balance between the abilities, needs, and expectations of an employee and the requirements of the environment and the possibilities for their fulfillment.

WORK EFFICIENCY the ratio of results achieved to resources expended, which means achieving the intended goals with the least possible consumption of resources.

WORK ENVIRONMENT the surroundings with which an employee comes into contact while performing their duties.

WORKS HYGIENE is a science that studies the impact of working environment conditions on employees' health and prevents the negative effects of work.

WORK SAFETY a set of conditions and technical, organizational, and economic measures in an organization that allow for the safe performance of tasks and the prevention of accidents at work.

WORKING CONDITIONS all factors present in an organization that are of specific importance to its employees.

Index

A

Adverse event, 69
Aggression, 38
  active, 40
  direct, 40
  electronic, 41
  hostile, 32, 38, 40
  indirect, 40
  instrumental, 40
  interpersonal (between people), 40
  passive, 40
  physical, 40
  psychological, 40
  theories, 35–36
  scale of the phenomenon, 47, 66, 107
  verbal, 40
Aggression (categories), 39–41
  online (electronic) aggression, 41
  patient aggression, 45–46
Aggressive behavior, 34, 41,88
  causes, 92
  effects, 92–94, 121
  indirect form, 40
  physical form, 40
  psychological (verbal) form, 40

B

Behavioral assessment room, 97
Blended approach, 117

C

Challenges, 3, 88
Change in the healthcare system, 99
Checklists, 98, 113
Classification of hazards at work, 17
Code grey, 98–99
Culture of occupational safety, 18, 95
Culture of safety, 18
  strong safety culture, 18–20
  weak safety culture, 20

D

De-escalation policy, 115

E

Emergency medical services, 6
Emergency medicine 5
Employee mutual support program, 98
Employee solidarity, 98
Ergonomics, 17
  corrective ergonomics, 17
  conception ergonomics, 17

H

Hazards in the workplace, 16
  biological agents, 26–27
  chemical factors, 27–28
  dangerous factors, 22–23
  harmful factors, 22–23
  intangible (social) factors, 16–17
  material factors, 17, 23
  nuisance factors, 22–23
  physical factors, 23–25
  psychosocial factors, 28–33

L

Letters to patients, 97
Limited quality control, 8

M

Mental strain, 16–17

N

NIOSH (National Institute for Occupational Safety and Health), 95

O

Occupational risk assessment, 17
Occupation risk assessment cards, 108
Occupational safety, 16
Occupational stress, 31–32
Organizational silence, 77
OSH management (occupational safety and health), 19–20

P

Patient's perspective, 46–48
PERMA model, 13
Physical strain, 7, 28–30
Prevention committee, 97
Psychological complexity, 8

Q

Quality of human resources management, 11

S

Safewards program, 98
SERS (Safety Event Reporting System), 96
Shift lag, 30
Shortage of healthcare workers, 12
SOAS-R (Staff Observation Aggression Scale – Revised) ix, (Foreword)
Stakeholder, 120–121
Strategies, 94–120
  comprehensive strategies, 110, 120
  first stage, 113–114, 120
  second stage, 115, 120
  third stage, 116, 120
Support strategy for victims, 108
Sustainable work environment, 13–17

T

Trauma counseling program, 97
Triage systems, 8

V

Violence, 38
  definitions, 38
  emotional, 39
  physical, 39
  psychological, 39
  violence preventing program, 96
  workplace violence, 42–43

W

Weapon detection system, 111– 112
Well-being, 13–16
Work environment, 11–12, 14
  intangible factors, 17, 23
  intangible work environment, 11,17, 23, 112
  material factors, 17, 23
  physical work environment, 111–112
Working time, 11